11-14-72

How To Know

THE

MAMMALS

Pictured Key Nature Series

How To Know

THE
MAMMALS

Ernest S. Booth

Anacortes, Washington

WM. C. BROWN COMPANY PUBLISHERS
Dubuque, Iowa

Copyright © 1950, 1961 by
H. E. Jaques
Copyright © 1971 by
Wm. C. Brown Company Publishers
Library of Congress Catalog Card Number: 79-129604
ISBN 0–697–04870–5 (Cloth)
ISBN 0–697–04871–3 (Paper)

THE PICTURED-KEY NATURE SERIES

How To Know The—

AQUATIC PLANTS, Prescott, 1969
BEETLES, Jaques, 1951
BUTTERFLIES, Ehrlich, 1961
CACTI, Dawson, 1963
EASTERN LAND SNAILS, Burch, 1962
ECONOMIC PLANTS, Jaques, 1948, 1958
FALL FLOWERS, Cuthbert, 1948
FRESHWATER ALGAE, Prescott, 1954, 1970
FRESHWATER FISHES, Eddy, 1957, 1969
GRASSES, Pohl, 1953, 1968
GRASSHOPPERS, Helfer, 1963
IMMATURE INSECTS, Chu, 1949
INSECTS, Jaques, 1947
LAND BIRDS, Jaques, 1947
LICHENS, Hale, 1969
LIVING THINGS, Jaques, 1946
MAMMALS, Booth, 1949, 1970
MARINE ISOPOD CRUSTACEANS, Schultz, 1969
MOSSES AND LIVERWORTS, Conard, 1944, 1956
PLANT FAMILIES, Jaques, 1948
POLLEN AND SPORES, Kapp, 1969
PROTOZOA, Jahn, 1949
ROCKS AND MINERALS, Helfer, 1970
SEAWEEDS, Dawson, 1956
SPIDERS, Kaston, 1952
SPRING FLOWERS, Cuthbert, 1943, 1949
TAPEWORMS, Schmidt, 1970
TREMATODES, Schell, 1970
TREES, Jaques, 1946
WATER BIRDS, Jaques-Ollivier, 1960
WEEDS, Jaques, 1959
WESTERN TREES, Baerg, 1955

Printed in United States of America

INTRODUCTION

AMMALS, the most intelligent and the most interesting of living creatures, command the attention of every person in the world, whether he has any interest in biology or not. He is, of course, a mammal himself, duly catalogued as phylum, *Chordata;* subphylum, *Vertebrata;* class, *Mammalia;* order, *Primata;* family, *Hominidae;* genus, *Homo;* species, *sapiens.* Mammals are separated from the rest of the animal kingdom by thirty or more characteristics, chief of these being the presence of hair or fur, and of mammary, or milk glands.

Under the same class, *Mammalia,* are such diverse creatures as the sea-going whale and the sea lion, the swift winged bat and the flying squirrel, the slow moving sloth, and the sly house mouse, and of course, all the zoo favorites from the monkey house to the elephant pen. Then there are the domestic animals who have aided man for thousands of years and who, even in this machine age, still have an important place.

1724853

It is somewhat surprising to discover that most people have seen only a very few kinds of the more than 10,000 species that make up the mammal population of the world. Mammals are so secretive in their habits that many of us go through life without knowing that twenty to fifty kinds of wild mammals live in our very neighborhood. True, most of these mammals are small folk, such as mice, gophers, bats, moles, rats, and shrews, but some of them are as large as a fox or coyote, and they all may live in populated areas where we might least expect wild mammals to occur.

If one had the urge to study mammals while traveling around the United States and Canada it would be possible to discover more than 1500 wild varieties of the furred creatures without going outside the boundaries of our country. It is with this wild population that we are particularly interested in this book. Domestic animals and some of the foreign species commonly seen in zoos have been included in the Key to the Orders so that you might better understand the family relationships of mammals. But by far the most numerous are the small mammals of woods and field. If one is interested enough to capture these animals, he will be able by the aid of the pictured-key to identify any he may find, be it the kangaroo rat of the western deserts or the lemming mouse of the north woods.

In compiling data for these keys, I have made use of hundreds of books, journals and papers—too numerous to name here. The most helpful of these to mammal students I have listed under Reference Works. However, one book needs to be mentioned in particular, for it stands above all others. It is THE MAMMALS OF NORTH AMERICA by E. Raymond Hall and K. R. Kelson. This is the first book ever to name, describe, and list the geographic range of every species and subspecies of mammal known from North America. This book revolutionized mammalogy, and at once became the final authority on names, descriptions and ranges. Everyone interested in mammals will at one time or another need to refer to this book.

I wish to express my appreciation for the excellent drawings used in my book, and made by my old friend and former student, Harry Baerg. Mr. Baerg has studied mammals for many years himself, and portrays a great many from personal observation.

Anacortes, Washington

Ernest S Booth

CONTENTS

THE SCIENCE OF MAMMALOGY

 AMMALOGY, or the study of mammals, has not always been a separate science. With the advance of knowledge in biological fields it became necessary for men to separate the study of animals and plants into smaller categories whereby individuals could follow a certain line of study much farther than would be possible if he had to master the whole field of biology. Mammalogy developed in America during the early part of the nineteenth century, with John James Audubon and John Bachman among the leading mammalogists of the time. Others, like Lewis and Clark, J. K. Townsend, and Charles Wilkes were early explorers who pressed westward making many of the discoveries of mammals in the unexplored parts of the United States. Toward the middle of the century the United States government sent out parties to survey the western states for possible railroad routes to the Pacific. Each party took naturalists with them to study the plants and animals. These men recorded many observations on the mammal life of the western states, describing dozens of kinds of animals that had never been known before.

Then, toward the end of the nineteenth century, the National Museum staff took over the study of mammals, organizing the Biological Survey. This group of men carried on an intensive study of all parts of the United States, under the leadership of C. H. Merriam. They formed the largest collection of mammals in the world, which at present, housed in the National Museum, has over 250,000 specimens in all.

In our day, the large universities such as California, Michigan, Harvard, and Kansas have outstanding courses in the study of mammals, while the main museums such as the American (New York), National (Washington, D. C.), Carnegie (Pittsburgh), and the Chicago Museum are noteworthy in their collections and literature on mammals.

But it is not necessary to attend one of these universities, nor to visit one of these museums in order to learn about mammals. Many will never have an opportunity to get beyond the fields and woodlands about their own homes in the study of animals. Even for such a one the science of mammalogy can become a very real hobby if he will take the time to learn a few methods for the study of mammals.

FOSSIL MAMMALS

TWO thirds of all the genera of mammals that ever lived on the earth are extinct! Now a genus (singular of genera) includes, as a rule, several closely related species. This means, then, that while we may have more new species and subspecies of mammals forming all the time, we have only one third as many genera represented today as have in all ages roamed our earth. One would suppose that we could go out into the fossil beds of America and find hosts of mammals buried in the ground, and that is exactly what we can do, for America is unusually favored by a large group of fossil mammals.

In the days when these animals inhabited the plains of the midwestern states, America must have looked more like Africa than any other part of the earth. In addition to our abundant antelope, bison, and deer were the rhinoceros, camels, giraffes, giant pigs, tapirs, sloths, mastodons, mammoths, and hundreds of others that do not even resemble the present day mammals. Many types of horses occurred in the western part of the country, some not more than twelve inches high, and with three toes instead of one large hoof.

Great deposits of fossil bones, mostly fragments, but many of them complete skeletons, can be found in such places as John Day, Oregon; Wind River, Green River, and Bridger, Wyoming; Rancho

La Brea, California; Uinta Basin, Utah; and Devils Gulch, Nebraska. These are only a few of the famous American fossil beds, but in these regions thousands of specimens have been found, and most of our museums contain many specimens of these prehistoric mammals. Much of Texas and Oklahoma, large areas in New Mexico, Colorado, Nebraska, and South Dakota are vast burial grounds for extinct mammals.

The study of fossil mammals is an interesting branch of mammalogy, but it is often disappointing to the beginner who goes out in search of specimens for himself. Most of his finds will be only fragmentary. One must become an expert in the study of bones before he can identify fossil mammals from such fragments. But his search can be worth while, for most of the large museums will identify fossil specimens for anyone who wishes to send them in.

Many giant mammals occurred in the past, and a few are sketched (fig. 1) to show the grotesque types of animals that used to inhabit the plains and forests of America. Some of these were so large that an elephant would appear to be a pygmy beside them. Many other animals

Figure 1. A few of the many extinct mammals. A, Horned Gopher, **Epigaulus hatcheri**; B, an Elephant-like Amblypod, **Uintatherium alticeps**; C, Giraffe-camel, **Alticamelus altus**; D, Giant Pig, **Archaeotherium ingens**; E, a Glyptodont, **Doedicurus clavicaudatus**; F, American Mastodon, **Mastodon americanus**; G, an early Horse; H, Sabre-toothed Tiger, **Smilodon californicus**; I, a Litopterna, **Macrauchenia patachonica.**

were very small. Besides *Eohippus*, the twelve-inch horse, there was a diminutive deer about the same size. If one is interested in further study of fossils I would refer him to the book by Scott (1937) listed with the references.

ADAPTATIONS OF MAMMALS

NE cannot study mammals without realizing that most of them are remarkably well adapted to their own type of environment. Or, could it be that the animal chose the environment because of his peculiar characteristics? In any case, we can find numerous examples today where the animal seems to be living just where it should be in order to use its body to the greatest advantage.

Take the gopher, for instance. This little fellow is built to dig in the ground. His front feet have very long claws that dig tunnels rapidly. He has cheek pouches so that when he comes to the surface of the ground he can run out from his tunnel a short distance, fill his face with tender shoots of grass, then rush back into his tunnel before a hawk or owl comes along. On some nights, not all gophers rush fast enough, for I have found nests of horned owls containing as many as five or six gophers. The gopher has very small eyes, and almost no external ears at all, although he can hear very well.

Figure 2

The mole (fig. 2), too, is well adapted to live in the ground. Because he almost never comes out of the ground, he escapes the hawks and owls. The mole has spade-like front feet, almost no eyes at all, and practically no ears. He has no cheek pouches, but he does not need them for he catches insects and earth worms down in the ground.

The tree-dwelling mammals such as the tree squirrels have special knobs on their toes that enable them to climb about in trees with great ease. A tree squirrel can run down a tree trunk head first, but a cat must back down the tree, for it does not have the toe pads. Then there are the gliding flying squirrels, with a thin membrane-like skin along the sides of the body which can be stretched out so that the animal is enabled to glide from one tree to another. Even the hairs on the tail grow longest on the sides, so that the tail is flat. The body of the flying squirrel is very light in weight for its size.

The only true flying mammals are the bats. They are among the animals most perfectly adapted for a particular environment. Their front legs are small, but their toes are very long, with a

thin membrane between each one forming the largest part of the wing. There is a hook or claw at the bend of each wing which enables the animals to cling to the rocky wall of a cave or to the rough rafter of a building. The tail also has a membrane on each side which serves as an insect net to trap tiny insects while it flies through the air. These are caught in the tail membrane, then the bat bends backwards and picks off the insect with his extremely sharp, needle-like teeth, and chews it up at once. Bats are able to fly rapidly at night through branches of trees, and into dark caves where they could not possibly see. An amazing form of radar guides them unerringly through the darkness. The bat emits a continuous series of high-pitched squeaks (out of our range of hearing) producing sound waves which strike objects as it flies along. The echo from these sounds is picked up by its highly sensitive ears, and enables the bat to know where there are obstructions to his flight. This "radar flying" makes a bat very hard to catch in an insect net.

Then there are aquatic mammals like the whales, porpoises, and sea cows or manatees. These animals are truly aquatic, for they never come out of the water willingly, and when washed onto a shore they soon die. Their limbs are not legs, but fin-like flippers, and these animals are expert swimmers, often resembling fish to the inexperienced observer. Other mammals are semiaquatic, such as the

Figure 3

seals and sea lions. They live in water most of the time, but may be found on the shores of rocky islands when they bear their young, or during migration. When on land these animals are very clumsy, for their main adaptation is for life in the water. Sea otters are now almost entirely aquatic, for they are able to bear their young, take care of them, and carry on all activities in the water (fig. 3). In former times they came out of the water to bear their young. Now, however, due to persecution by man, they have resorted completely to the water to escape extermination. This is very unusual, for ordinarily we would say that man cannot change the habits of wild animals.

A study of the teeth of mammals is interesting.   Teeth of cattle, deer, elk, and antelope are efficient grinding organs, well adapted to an herbivorous diet of grass and green plants.   On the other hand, teeth of carnivorous mammals such as the mountain lion are adapted to the tearing of meat.   The teeth of beaver show a remarkable adaptation to their work of cutting into wood.   In addition to their size and strength they are self-sharpening.   They are formed with a thick enamel layer on the outer surface, and a layer of softer dentine on the inner surface.   As the beaver works away on the trees, the dentine wears off, leaving a sharp cutting edge of enamel.   The four front teeth grow constantly, otherwise they would soon be worn down to uselessness.

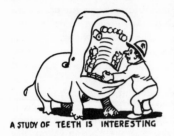

A STUDY OF TEETH IS INTERESTING

Yet, if one tooth is broken the opposite one in the other jaw will not have a mate to grind against, and will continue to grow without being worn down.   Cases are on record where teeth have grown so long that they stuck out of the mouth of the animal in such a fashion that it could hardly eat.   A few cases are recorded where the teeth have actually grown around outside the mouth and back into the brain, thus killing the animal.

One of the most spectacular phases of adaptation is found in the coloration of mammals.   Unlike the birds, sex plays very little part in color variations of mammals.   Rather, they vary according to the soil, climate, time of year, and age.   The pocket gopher, for instance, seems to depend on the soil color for his particular shade. In pale colored sandy soils the gophers are light in color, while in dark colored soils the fur is dark.

Mammals in damp climates are almost always darker in color than those in arid regions.   This variation in color is one of the chief points for distinguishing subspecies of mountain lions, mice, shrews, bats, and other animals. Many mammals change their coat in summer and again in winter, and the colors are usually different.   The most unusual change of this type is found in the snowshoe rabbits and in the weasels.   These animals are brown or grayish brown in summer and pure white in winter, except for black tips on the ears of the rabbits, and a

I SEE MRS. LEPUS IS SPORTING A NEW WINTER COAT

ze anything has happened, and on larger forms such as
rabbits, ear punching or marking with numbered metal
the legs might be more advisable.

records of captures must be kept, and all these plotted
:ale map, so that the exact ranges of the animals can be
After several weeks of continued study in a small area
possible to plot the range of any kind of mouse or shrew
:ative accuracy.

ons are more difficult to study, for this kind of work must
on over a period of years. Biologists have discovered
ulations of certain kinds of mammals will, in a given area,
a high number, then suddenly it will decrease to a very
r. After this, the population usually begins to increase
is we have definite population cycles. Only a little study
en given to this phase of mammalogy. Serious mammal
e urged to carry on population studies.

.ada the populations of the lynx and the snowshoe rabbit
.ated periodically through the years. When the rabbits in-
numbers, the lynx also increased. Then when the rabbits
decline in numbers, the lynx also declined sharply in num-
 must not be mere coincidence, for comparative records
made for more than half a century, compiled from trap-
:ds. Evidently the number of lynx in an area depends
.umber of rabbits available for food. If food is scarce the
die, reducing the population to smaller numbers. More
the fact that during lean years the average number of
to a single female is less than during good years.

are unusual cases recorded of meadow mice increasing
.tly in an area, forming such a large population that mass
became necessary. On several occasions hundreds of
or even millions of mice migrated across the country to

MICE MIGRATE ACROSS THE COUNTRY

another area. These migrations have
also been observed in tree squirrel
populations, and they occur almost
regularly in lemmings in the arctic.
Practically nothing is known as to
the "whys" of these mass migrations, or even as to why populations
as violently as they do. Some have thought that disease
 population when it gets too large, but disease could hardly
regularly as the snowshoe rabbit fluctuations in Canada.

black tip at the end of the tail in the weasels. It is worth noting
that individuals in the same species of weasel or rabbit living in
regions where there is seldom snow in the winter do not change to
the white coat.

ECONOMIC IMPORTANCE OF MAMMALS

HEN one considers the economic importance of mammals
he may first think of domestic animals. Such animals,
most of which are not native to our country, have con-
tributed very much to the development of our present
civilization, and without them our economy would be
greatly impaired. Food products would rank as their
greatest contribution, then perhaps hides and furs, and medicinal pre-
parations made from glands and other parts of animals. The work
they do is not now so vital to American economy, but it is still of
great importance in less mechanized countries.

Fur bearing animals contribute millions of dollars worth of furs
each year. Formerly, before the killing of sea otters was banned

Figure 4

by law, one pelt would bring as
much as two thousand dollars. Nor
have the mammals of by gone ages
lost their commercial value. Fossil
mammoths, imbedded in arctic ice,
still contribute ivory (fig. 4). The
whale, with its great supply of oil,
is of economic importance.

Insect-eating mammals such as bats
and shrews are among the very important friends of man, for they destroy countless hordes of pests every
year. In some parts of the country large bat roosts have been built
to encourage bats to inhabit the area. Bats are perhaps the most
important as a control measure for mosquitoes, and as such they are
certainly worth their weight in gold.

Moles and gophers gain some small credit by aerating the soil.
Gophers in the mountains cause valuable soil to be washed into streams,
so that it is carried down and deposited in the valleys. This practice
might not be looked upon with favor by the soil conservationist, for
if it were carried to extremes it could be the cause of erosion and floods.

Wild game animals, as such, have economic value. And no one would deny the value of the wild animals of parks and recreation areas as tourist attraction.

Not all economic interest in mammals is on the positive side; a great many animals are harmful and destructive. The most serious pests to the farmer are gophers and many kinds of mice and rats, including, unfortunately, the imported European rats and mice—the most destructive of all. The wild mice that live in the fields cause damage to young plants, and make tiny trails through them. Some kinds such as the harvest mice actually climb up the wheat stalks and eat the grains from the heads. The European mice and rats cause their most serious destruction to stored products in warehouses and granaries. Rabbits, beavers, and muskrats may, in some places, be destructive. Beavers may cause damage to irrigation ditches, and may cut down too many trees.

Small carnivores such as weasels and skunks are fond of poultry, while coyotes, foxes, mountain lions, and bears may do great damage to cattle and sheep. Most of these animals can be effectively controlled by government trained predatory hunters.

The most serious indictment against mammals is the transmitting of diseases to man and domestic animals. There are a great many which are guilty. Rats head the list, for they are responsible for carrying the fleas which transmit bubonic plague, a disease which has been responsible for millions of human deaths during past ages. Nor is bubonic plague extinct today. In a recent survey in the Yakima Valley of eastern Washington, bubonic plague was discovered to be present among a group of seemingly harmless meadow mice living out on barren hills some distance from the nearest town or populated area. Ground squirrels and rabbits are known to carry tularemia, another disease somewhat similar to bubonic plague, but much less fatal.

Deer and rabbits are aids in the transmission of spotted fever, for they harbor the ticks which carry the disease. While these ticks do not bite man, they transmit the disease to the animals, who again may be bitten by another tick, *Dermacentor andersoni*. This tick has no scruples against biting humans, and thus tick fever is carried.

Still other diseases like typhus, and a host of parasites such as tapeworms, roundworms, and other internal parasites may be contracted from wild or domestic animals. Some of these diseases may be readily treated; others, like trichinosis (a parasite found in improperly cooked pork) can never be eradicated from the body, and may often be fatal.

HOME RANGES,
POPULAT

VERY mammal du
home range. W
it leaves the hor
own, usually not
it establishes its
travels in getting
this is the center of the home r
the extreme limit of the range u
mammal, and upon its feeding
range of several miles, while a
in an area not exceeding fifty fe

Many mammals have a perm
throughout their lives. Others h
they inhabit only during the bree
nature, for they are usually wande

Migratory mammals will have
entirely different range in the wint
migratory mammals return to the
many kinds of birds do return.

If a mammal protects its home
range, driving away intruders, especially of the same species, we
say that the animal has a territory. In other words, it seems
to own its home range in much
the same way that people will
own land or property. Territories may be established around a co
ing site.

The study of home ranges and
can be undertaken by anyone. The
traps, marked, released, then trappe
such trapping is carried on rather ex
soon discover how far any certain an
ing site. Marking of the specimens m
A poultry punch may be used, and sr
—if they are large enough. The syst
numbers by cutting off various toes.
punching and toe clipping to number u
clipping is most practical on small mam

seem to re
muskrats or
bands abou

Careful
on a large-
determined.
it would be
with comp

Popula
be carried
that the po
increase t
low numb
again. Th
has ever
students a

In C
have fluct
creased in
suffered a
bers. Th
have beer
pers' rece
upon the
lynx will
strange i
lynx borr

Ther
very gre
migratior
thousand

MILLIONS

fluctuate
attacks
recur sc

The study of populations can be undertaken by anyone who has patience enough to take censuses. A population census can be made for small mammals in a limited area by trapping it carefully and thoroughly for a short time. This will give a somewhat accurate count of the total population in the area. If you want to keep the animals in the area, then live traps will be necessary, and marking of specimens will be required. Large mammals may be counted directly from an airplane, or by having several helpers walk systematically over a given area. Such censuses cannot be entirely accurate, but will certainly serve the purpose. Censuses for carnivores may be difficult to secure, and often only rough estimates can be made. This difficulty, and the time and patience required to make even the easiest census, are reasons why we know so little about mammal populations.

REFERENCE BOOKS OF GENERAL INTEREST

Blair, W. Frank
   Vertebrates of the United States. (New York: McGraw-Hill, 1957.)
Burt, William H.
   A Field Guide to the Mammals. (Boston: Houghton Mifflin, 1952.)
Cahalane, Victor H.
   Mammals of North America. (New York: Macmillan, 1947).

   _____,
   Meeting the Mammals. (New York: Macmillan, 1943.)
Cockrum, E. Lendell
   Introduction to Mammalogy. (New York: Ronald Press, 1962.)
Davis, David E. and Frank B. Golley
   Principles in Mammalogy. (New York: Reinhold Publ. Corp., 1963.)
Grosvenor, Melville B. et. al.
   Wild Animals of North America. (Washington, D.C.: National Geographic, 1960.)
Hall, E. Raymond, and Keith R. Kelson
   The Mammals of North America, 2 vols. (New York: Ronald Press, 1959.)
Hamilton, W. J.
   American Mammals. (New York: McGraw-Hill, 1939.)
Pike, Oliver G.
   Nature and Camera. (New York: The Focal Press, 1943.)
Rue, Leonard Lee III
   Pictorial Guide to the Mammals of North America. (New York: Thomas Crowell, 1967.)
Scheffer, Victor B.
   Seals, Sea Lions and Walruses. (Stanford, Cal.: Stanford Univ. Press, 1958.)
Scott, William B.
   A History of the Land Mammals of the Western Hemisphere. (New York: Hafner, 1962.)

Walker, Ernest P. et. al.
Mammals of the World, 2 vols. (Baltimore, Md.: Johns Hopkins, 1968.)

REFERENCE BOOKS OF RESTRICTED AREAS

Bailey, J. W.
The Mammals of Virginia. (Richmond, Virginia: The Author, 1946.)

Bailey, Vernon.
Mammals of New Mexico. (Washington, D. C.: U. S. Government Printing Office, 1931.)

——————.

Mammals of North Dakota. (Washington, D. C.: U. S. Government Printing Office, 1926.)

——————.

The Mammals and Life Zones of Oregon. (Washington, D. C.: U. S. Government Printing Office, 1936.

Berry, William D. and Elizabeth
Mammals of the San Francisco Bay Region. (Berkeley, Cal.: Univ. of California Press, 1959.)

Booth, Ernest S.
Mammals of Southern California. (Berkeley, Cal.: Univ. of California Press, 1968.)

Burt, W. H.
The Mammals of Michigan. (Ann Arbor: University of Michigan Press, 1946.)

Dalquest, Walter W.
Mammals of Washington. (Lawrence, Kansas: Univ. of Kansas, 1948.)

Davis, W. B.
The Recent Mammals of Idaho. (Caldwell, Idaho: Caxton Press, 1939.)

Goodwin, George G.
Mammals of Connecticut. (Hartford, Connecticut: State Geological and Natural History Survey, 1935.)

Hall, E. R.
Mammals of Nevada. (Berkeley, California: University of California Press, 1946.)

Hamilton, W. J.
The Mammals of Eastern United States. (Ithaca, New York: Comstock, 1943.)

Handley, C. O., and C. P. Patton.
Wild Mammals of Virginia. (Richmond, Virginia: Commission of Game, 1947.)

Howell, Arthur H.
Mammals of Alabama. (Washington, D. C.: U. S. Government Printing Office, 1921.)

Ingles, Lloyd G.
  Mammals of the Pacific States, California, Oregon, Washington.
  (Stanford, Cal.: Stanford University Press, 1965.)
Peterson, Randolph L.
  The Mammals of Eastern Canada. (Toronto, Ontario: Oxford University Press, 1966.)
Swanson, Gustav, T. Surber, and T. S. Roberts.
  The Mammals of Minnesota. (Minneapolis: Minnesota Department of Conservation, 1945.)
Warren, E. R.
  The Mammals of Colorado. (Norman, Oklahoma: University of Oklahoma Press, 1942.)

HOW TO STUDY MAMMALS

 HE study of mammals entails a great deal of work on the part of the one who would like to learn about mammals from first hand experience. It is one thing to obtain book knowledge about mammals, and quite another to get real field experience. It is the hard way we are describing in this book. The references provided will be helpful as supplementary information. Before beginning our study we might do well to consider the equipment we will need, the methods for using the equipment, the process of skinning and preparing specimens, and ways to identify the specimens once we get them.

But first ——

TO TRAP OR NOT TO TRAP

There are people we meet now and then who think it wrong to allow students to collect mammals, and to prepare them as museum specimens. There are even societies which attempt to prevent the killing of animals for study purposes. It is true that collecting mammals entails the use of traps, and some animals are sure to be hurt in traps. But those who condemn the taking of study specimens would do well to consider all the facts before judging the biologist too harshly.

In the first place, will it cause a person to become cruel if he sets mouse traps out in the woods anymore than if he sets them in the basement or on the pantry shelf? The housewife feels great disgust toward the mouse eating the cheese in her cupboard. Should she then condemn the mammalogist who wants to learn about the mice out in the forests or in the deserts? Recall, too, that many of the little animals are actually harmful to man. No one would lose sleep over the killing of a weasel which catches his chickens. Then should the mammalogist be condemned who catches a chipmunk on the mountain side? True, the chipmunk does no harm up there, but neither does the weasel up there. If the chipmunk lived about the house it would certainly do some damage.

No one condemns the fisherman or the hunter, and they kill not in self defense, nor out of necessity for food. Actually, the average hunter pays many times more for his game than an equal amount of meat would cost from the grocery store. We maintain, then, that the student of mammalogy should not be condemned because he wants to learn more about mammals. It is true that he will have to kill some of them in learning, but he can certainly learn to kill them humanely, and he will certainly not kill more than he needs. In many cases he can use live traps, so that the animal will not suffer pain while in the trap. Several such traps are described in this book.

EQUIPMENT

Some equipment will be necessary in the study of mammals. The following list is only suggestive, for some items will not be absolutely necessary, while a serious student will want even more than this list includes.

25 or more mouse traps (Victor 4-Way are best).

6 to 10 rat traps (Victor 4-Way).

2 to 10 No. 1 steel traps, if fur-bearing mammals or larger kinds like woodchucks are to be taken.

2 to 10 mole and gopher traps. There are several types of these on the market which are satisfactory; the spear-type mole trap is the best.

1 or more carrying bags for traps, bait, and other equipment.

1 shotgun, .410 or 20 gauge is best.

1 or more boxes of shotgun shells, size 9, 10, or 12 shot.

1 carrying case with handle, for skinning equipment.

1 skinning kit, made up as follows:

    1 fine pointed scissors

    1 heavier scissors for cutting small bones

    1 scalpel or pocket knife with high grade steel

    1 rule with millimeter scale

    1 forceps, 6 inches, fine tips

    1 package needles

    1 package pins (glass headed pins feel better to the fingers)

    1 spool No. 8 thread, for labels and for sewing large mammals

    1 spool No. 20 or 24 thread, for sewing small mammals

    1 spool annealed wire, No. 24 for mice

    1 spool annealed wire, No. 20 for chipmunk-sized mammals

    2 or 3 sizes of larger wire for large specimens

    1 tooth brush for smoothing out fur

    A carborundum stone for sharpening tools

    1 small roll absorbent cotton for stuffing tails and legs

1 large roll quilting cotton for bodies

Excelsior for centers of bodies of large mammals

Vernier calipers for measuring skulls, ears, feet, etc.

1 pint commercial formalin (40 percent).

1 pint ether or chloroform if animals are to be killed after being caught in live traps.

Labels—these may have your name printed at the top.

1 bottle Higgins India ink with fine-tipped pen (not a fountain pen.)

2 notebooks for records; one for field catalogue, the other for field notes.

1 copy of state game laws.

Permits and licenses for collecting specimens. Some states require no permits for small mammals, but all do for fur-bearing mammals. *No specimens should be collected until you know you are taking them within your rights as a citizen.* Write to the state game department for information.

COLLECTING SPECIMENS

This phase of mammalogy is one which takes a great deal of time and outdoor activity if it is to become a vital part of your study of mammals. First, you must know something about the habits of the mammals you want to study. Books will help, but don't forget that as you get into this work you will soon be absorbing much mammal lore right out in the field. If you come upon an animal going about its business without knowing of your presence, do not be in a hurry to collect it as a specimen. Sit still and watch it as long as you can. Every minute of observation while it is alive in its native surroundings will be worth much more to you than the stuffed skin in your collection.

But sooner or later you must resort to trapping or shooting specimens in order to obtain one or more of each kind in your locality. If you do not want to keep the animal alive for study, then traps that kill the animal on the spot will be best. The ordinary mouse traps and rat traps will do very well for most of the small animals. They are usually set at the entrance hole of the nest, or placed in a runway. Here, one must learn a bit about the lives of the creatures he is studying in order to capture them successfully.

If you want meadow mice you will need to find small tunnel-like runways in thick grass. Set the mouse traps across the runways with the trigger of the trap in the center of the runway. No bait will be needed, for the mouse merely runs over the trap and is caught. Other mice, like the deer mouse or white-footed mouse (*Peromyscus*) will be caught in mouse traps set under logs and about the base of trees or stumps. The bait is rolled oats sprinkled on the trigger end of the trap. Do not be afraid of using too much bait for these wild mice love to store food, and if you put plenty of bait on the trap the creature will be back to get all of it even if it takes him half the night. Remember that you will seldom catch anything in the day time, so set the traps late in the afternoon, and pick them up early in the morning. It is not best to leave them set during the day, for birds often find them, and are killed.

Some mammals require special trapping technique. I shall list a few of these.

SHREWS.—Shrews can be caught in mouse traps set along streams or in marshy places. Rolled oats are usually sufficient bait, but some mammalogists claim that peanut butter mixed with the oats is even better. Shrews nest under logs and in old rotten stumps, so traps set

Figure 5

near these may be fairly sure of taking a specimen or two. I have had good success burying a milk bottle or deep tin can in the soft ground near a stream, than hanging a small piece of meat on the end of a string suspended from a stick projecting up over the opening of the can or bottle (fig. 5). As the shrew jumps up to catch hold of the meat, he falls into the bottle and cannot get out. If you want to catch several in one such trap just put water in the bottom so the animal will drown. If you do not drown each one you will find only one remaining in the morning, for one of the fierce little fellows will eat up the others as they fall in. The tin can mouse trap described later will be quite effective for catching shews. Shrews need so much food to keep them alive, however, that often they will have starved to death by the time the trapper finds them the next morning. If you want live shrews it would be best to run your trap line several times during the night, and to remove every shrew while it is still alive.

Shrews must have constant care in captivity, for they eat more than twice their own weight every day and must have a constant supply of food. No one has ever kept a shrew in captivity longer than a month or so at the most, and no one has ever succeeded in raising them in captivity. I believe it could be done if one had time and patience to catch enough mice and insects to keep the shrews alive. A shrew will kill a mouse bigger than itself, and eat most of it on the spot.

MOLES.—Moles can be caught best in special mole traps, although ordinary gopher traps will do. All mole and gopher traps must be placed down in the underground tunnels of the animals, and should be concealed so that the animals will not know anything is disturbed. A live trap can be made for moles quite easily, but it takes a lot of patience to use it, for the moles have an uncanny knack of avoiding it.

Moles have seldom been kept in captivity, yet one could make some interesting observations of them if they were kept in a cage filled with dirt, and provided with plenty of insect grubs and earthworms. A cage could be made from strong wood with glass sides, and filled with dirt. Since moles are blind they would often come to the glass surface with their tunnels, and one could easily see what the animal was doing.

BATS—Bats are very hard to get because they hide in caves, and fly only at night. One may occasionally get them at dusk with a shotgun, but fine shot is necessary for bats are delicate creatures. The best way to hunt bats is to pursue them to their hiding places with a butterfly net. It is sometimes possible to find clusters of from twenty to thirty bats hanging from rocky walls in a cave. Occasionally one may catch them at night by putting up a big sheet in the woods and turning the car lights on the sheet. As insects are attracted to the sheet the bats will come, too, and if one has a very large net he may catch a few bats. Old buildings, especially if they are deserted, often harbor bats. One had better be prepared with both gun and net to collect the agile creatures.

Bats can be kept in captivity by feeding them a constant supply of insects. Soon the bats will learn to take insects from the tips of forceps, and many interesting observations can be made from captive specimens.

RABBITS.—Rabbits are hard to trap, but they can be caught in the old figure-4 trap by arranging a box on sticks so that a tug on the apple placed on a stick will let the box fall down. The best luck collecting rabbits will come by use of the shotgun.

MICE.—Mice can be caught very well in ordinary Victor 4-Way mouse traps. These may be set in runways or at the entrances of burrows in the ground, with rolled oats for bait. You will need to learn what kinds of mice are in your locality and something of their habits so that you will know where to set your traps. White-footed mice (*Peromyscus*) often live in brushy places; meadow mice (*Microtus*) usually live in thick grass; harvest mice (*Reithrodontomys*) live in weeds along roadsides or in wheat fields; pocket mice (*Perognathus*) live in wheat fields and weedy places; jumping mice (*Zapus* and *Napeozapus*) usually live near water; red-backed mice (*Clethrionomys*) live in the deep forests of the mountains; phenacomys (*Phenacomys*) live in high mountain meadows. These are only a few of the wild mice, but this will give you some idea of their variety and the places where you may set traps. Live traps will be necessary if you want to keep them in captivity.

A simple live trap is made from a mouse trap and a tin can (fig. 6). The tin can is fastened to the mouse trap in such a way that the trigger is inside the can, and a piece of tin is placed on the

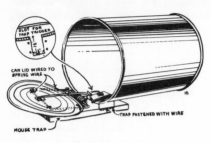

Figure 6

spring wire. When the trap snaps the tin comes over the opening of the can and makes a closed can with the mouse inside. A No. 3 fruit juice can is best for size. A large wad of cotton should be placed in the back of the can so that the mouse can make a nest for himself. This is very essential if the weather is cold. A slot must be cut from the can to make a place for the trigger, also a hole must be made in the lid to allow for trigger wire to stick through. The drawings will give the rest of the details.

RATS, SQUIRRELS, CHIPMUNKS.—Rats will be trapped best at night, but squirrels and chipmunks are out in the day time. These larger rodents may be caught in rat traps if they are not wanted as live specimens, although many squirrels are too big for rat traps, and steel traps must be used in that case. I recommend live traps for the larger squirrels, for steel traps are cruel things. Anyone can make a good box trap if he uses a little ingenuity (fig. 7). The following instructions may serve as a guide for those who have never made one.

Remove *both top and bottom* from a box such as an apple box, and examine each end to see where it will be best to make a cut 5 inches wide and 6 inches deep into the middle of one of the ends. Cut the corners of a piece of galvanized sheet metal 16 by 25 inches, and bend it to fit tightly around the open bottom of the box to form the floor of the trap. If the box is a standard apple box, there will

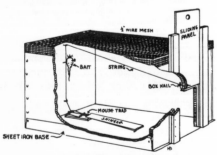

Figure 7

be 2 inches of metal overlapping the sides. If the box is not this size, measure the sheet metal accordingly. The overlapping edges should be nailed into the corners of the end boards to fit against all sides except the front end. This will strengthen the entire trap. Next place a piece of galvanized mesh wire netting, ½ inch (hardware cloth), size 16 by 22 inches, on top of the box with 2 inches overlapping the sides except the front. Small staples may be used to tack this into position. The wire netting permits light to enter the trap, making visible the bait to the animal, and making visible the animal to you—this would be a factor in case you captured a skunk!

The opening in the bottom of the front end will be closed by a sliding panel ⅝ by 6 by 14 inches when the trap is sprung. Two 12 inch strips of wood, ¾ by 2 inches, should be nailed vertically ⅝ inch from each side of the opening, and a ¾ by 3 by 12 inch strip nailed on top of the first strips with the outside edges of both strips flush together. This will form a sturdy casing for the sliding panel, and, when all the boards are planed smooth, and waxed, the panel will drop shut with ease when the trap is sprung. The sliding panel should be raised in its casing until it is about ¼ inch higher than the top of the opening and tacked there temporarily while a hole large enough for an 8-penney box nail is bored through both the sliding panel and the front end of the trap. The 8-penney nail can slide back and forth through the continuous hole in the trap end and the sliding panel, thus dropping the panel when the nail is pulled out by the snapping of the mouse trap.

A mouse trap furnishes a trigger release, and the bait is tied to the wire netting at the back of the trap just over the mouse trap. Fasten a cord to the head of the nail, which is just barely supporting the weight of the panel. The cord is drawn tightly to the back of the trap and looped over the strands of the wire netting, then tied to the snap wire of the mouse trap. The mouse trap is fastened in position at the back of the trap with the trigger facing the entrance hole. The cord attached to the mouse trap should pass through a small staple fastened just above the mouse trap on the back end of the box. Place a piece of shingle about 3 by 6 inches with the thin end resting on the trigger of the mouse trap. When the animal comes in and reaches for the bait it will step on the shingle, thus setting off the mouse trap, which in turn pulls out the nail, releasing the sliding panel.

This trap may be used for ground squirrels, chipmunks, rabbits, and even weasels and skunks. For the carnivores canned salmon or other meat will make good bait, while rabbits and squirrels prefer carrots, nuts, or apples. Salmon need be merely smeared on the shingle when it is used.

MUSKRATS, BEAVERS.—These animals are fur bearers, and must not be trapped except under trapping licenses. Commercial traps will work very well, but live traps are to be recommended. A very fine box trap made of metal can be secured from the F. C. Taylor

Fur Company, St. Louis, Missouri. This will also work very well for squirrels, rabbits, and other medium sized animals.

GOPHERS.—Effective gopher traps may be purchased at any hardware store. Most of these traps punch two holes in the sides of the animals, causing the fur to become rather bloody. This can be cleaned off with a little cold water. A live trap may be constructed that will work equally well for gophers and moles.

The trap is a long rectangular wooden box with hinged metal doors that swing in, but not out (fig. 8). The inside width and height should be about 4 or 5 inches. The ends of the side boards should be

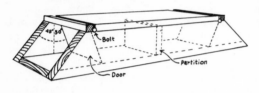

Figure 8

cut diagonally projecting several inches farther on the bottom edge so as to make a less abrupt approach to the suspended door. The edges of the top board should overlap the top edges of the side boards, and the edges of the bottom board should be overlapped by the side boards for strength and durability. Boards of ¾ inch thickness should be used in making this trap. The top board is 6½ inches wide, the bottom board 5 inches wide, the side boards 5¾ inches wide, and the length of the trap about 24 inches. The end of the top board should lack an inch of being flush with the ends of the sides in order that a ¾ by ¾ inch strip may be nailed over the end gap between the sides with a ¼ inch between the strip and the end of the top board. The metal door may be suspended from this empty hinge space. A 5 by 6½ inch piece of galvanized metal may be used for the metal door. The ½ inch end of the door may be wrapped around a ¼ by 7 inch bolt to act as a hinge. The metal door will then hang inward at an angle.

This arrangement will allow the gopher to push into the trap, but it cannot get out. It is best to make the other end of the trap exactly the same, so that the animal may be caught from either direction. In that case a heavy partition should be placed in the middle so that there would be no chance of two gophers getting in there and fighting. The animals may be removed from the trap by tipping the trap up over an open cage, then removing the door through the slot.

Gophers make their tunnels underground about 8 to 16 inches below the surface, coming up at intervals to push up excess dirt. Be sure to set the trap down in the main runway, not into a branch where dirt is pushed up, for the trap should be somewhat level in order to work properly. Dig out the main tunnel with a shovel, place the trap in position, and try to make the trap look as nearly like the original tunnel as possible by replacing dirt around the ends of the trap. Covering the whole thing first with boards so that dirt will not fall down on the trap entrance, place dirt over all to keep out light. Be sure to mark the place, for you might forget where the trap is.

CARNIVORES. — The box trap described for rats and squirrels will work very well for the smaller carnivores. Large animals like coyotes, foxes, bob cats, and bears will have to be killed with a rifle, or caught in large steel traps. Such specimens are best secured from government hunters and trappers, if you want them in your collection.

DEER.—Specimens of deer must be taken by rifle, or better yet, secure a head for mounting from some hunter who does not want his trophy. This will require much less effort and expense.

MARKING TRAPS

Students often find that they cannot remember just where they set many of their traps. The trap line may be marked by wrapping small bits of cotton about twigs or bushes near the trap area. A disadvantage of this method is that it attracts small boys who may be in the vicinity, and they may find the traps and move them, or take them away. I have found the following method satisfactory: set the traps in groups of three or five near each other, then move on to the next spot and make another set of the same number, and so forth. You will thus know how many you have in a spot, and it is not so hard to remember a few landmarks to mark the trapping areas. Until one gets used to mental plottings, it may be advisable to make a rough map of the area, indicating the various locations of your trap groups.

MAMMAL TRIPS

After you have trapped about your own home, you will want to take expeditions into new territory. It will surprise you how many new kinds of mammals you will find even on a short trip. A collecting trip may be very interesting, especially if you can go to some region where you have never been before. But for the true mammalogist, it is not the scenery, but the lure of new kinds of animals that keeps him going for days and weeks, so long as money for gasoline and food holds out.

Careful planning must go before the trip, for nothing must be left home that cannot be bought along the way. Besides the usual line of equipment, it would be wise to make a collecting case of some kind so that specimens may be pinned out in a screened cage, protected from flies until they are dry. Fly larvae will do great damage to larger specimens; small specimens generally dry before the fly eggs have time to hatch. Damage is greatest around the feet, nose, and the incision of the belly in larger specimens. Skulls may be badly damaged by fly larvae; they will become discolored, and may be made completely unpalatable to the dermestid beetles who will later clean the skulls. Frequent spraying of specimens with fly spray, especially with one using DDT, is helpful if a screened cage is not available. Do not spray skulls with DDT.

In late afternoon find a suitable place for camping and for trapping. Set out the traps before you make camp or prepare supper, for it is important to set traps by daylight if at all possible. Traps may be set by flashlight or gasoline lantern, but much better sets can be made in day light. After this is attended to, camp can be made. Early in the morning, the trap lines must be run, and skinning begun at once. Mammals must always be skinned during the morning for best results, especially in warm weather. Even then, some may spoil before they can be skinned. These can be made into skeletons, however, so they need not be wasted.

It is possible, after some practice and experience, for a person to trap and prepare twenty or more specimens in a day, and still have time left for camp duties. I would ask no better vacation than a trip through several states collecting mammals, even if I were to make twenty specimens a day for a month. This might get tiresome in time, but if you are travelling you will find so many new mammals that interest will always be keen.

Much of the interest in a mammal collection lies in acquiring as large a number of different kinds as possible. Mammal collectors all over the country are eager to trade specimens, providing your skins are standard quality. Anyone interested in trading specimens may have his name included in the *Naturalists' Directory*, published by The Casino Press in Salem, Massachusetts. The Directory includes names of mammal collectors all over the world to whom one may write.

PREPARATION OF SPECIMENS

SKINS AND SKULLS

 HE usual way to prepare a mammal for preservation is to skin it, make and insert an artificial body of cotton, sew it back together, pin it down on a board, attach the label, let it dry, and place it in a cabinet along with the cleaned skull, clearly labelled as to collection data, measurements, and other essential information. This is then spoken of as a study skin and skull. All the large museums have collections prepared this way, even though the general public never sees them. They are not suitable for display, but are ideal for study since they take up very little space, and can be handled easily and safely. This is the method to be described in this section.

SKELETONS

Too little attention has been directed to the making of mammal skeletons. Consequently, many large collections are lacking in these, when they just as well might have them. If collectors would take a little more time to save those specimens which are otherwise thrown away due to spoilage, many skeletons would be preserved. These need only to be "roughed out," dried away from flies, then cleaned by beetles just as the skulls are cleaned. "Roughing out" is merely skinning and removing the internal organs, and any larger bits of flesh that will come off readily, then allowing the animal to dry without decaying. When dry, it may be cleaned by beetles, then preserved in a cardboard box along with the same data which is saved for the skins and skulls. Skeletons of the animals which are prepared into skins should be saved as well, if the specimen is something unusual. Museums are generally desirous of obtaining skeletons of even smaller mammals for comparative study.

PRESERVING IN SOLUTIONS

It is possible to preserve mammals in solutions of several kinds. Bats may be preserved in 70 to 85% alcohol. While this method has been used a great deal in large museums, it is not so good as the study skin method. If a large number of bats are caught and it is impossible to skin all of them, the rest should be preserved in 85% alcohol.

On intensive collecting trips it is sometimes impossible to stuff all the specimens caught. There is a method whereby the animals may be preserved temporarily until after the trip is over, then skinned at a later time. This process is simple enough to use. The animal is injected in several places throughout the soft parts of the body with embalming fluid (formalin, full strength, 5 parts; phenol, full strength, melted crystals, 5 parts; glycerin, 5 parts; water 85 parts), then the entire animal is placed in the fluid. Data must be recorded as usual, and placed on a water proof label. In one or two months the animal may still be skinned and made into a good study skin. There will be much difficulty in cleaning the skull, however. At best, this method is an emergency method to be used only when specimens could not otherwise be saved.

DATA TO RECORD

It is hard for a beginner to understand that a specimen has no scientific value without a label. Record keeping is one of the most

particular phases of mammalogy. It is impossible to record too much
data about the specimen, and most people do not record enough.
First a label must be attached to the right leg of the specimen,
tied ¾ of an inch from the leg by No. 8 thread. The label may be
rather small, ½ inch by 2½ inches, or it may be as large as ¾ inch by 3
inches. If you can afford it, it is best to have your name printed at
the top, and your permanent address, if you have one. As illus-
trated (fig. 9), the trapping locality—state, county, and city—comes
next. If the place is not close to a city or town, it is best to record
the distance and direction from the nearest town. Any place that
can be found on a map, such as a lake or a mountain, is a suitable
locality. The next line is for the date; the third line for your name
(unless printed labels are used); the fourth line is for the measurements of the animal in millimeters. The sex (the sign ♂ symbolizes
the male, and ♀ the female) is indicated at the right end, and your
collector's number (beginning with 1 for your first specimen, 2 for
your second, and so on through life) after your name. On the back
of the label the scientific name comes first, then common name, if
any, and other information like habitat, elevation, or any other information you wish to record. One item that many leave off, but which
should always be recorded, is whether or not a female has embryos or

COLLECTION OF ERNEST S. BOOTH
WASH., PIERCE COUNTY, LONGMIRE
MAY 10, 1948
Ernest S. Booth –100   ♀
110 – 71 –20 –10   3 emb. 18 mm.

Figure 9

whether or not it is lactating. This is usually indicated on the front
side at the lower right hand corner.

Labels must be tied in a certain manner. Two holes are punched
in the left end of the label, and the thread passed through these.
The ends are crossed and passed back through the loop so that the
finished label looks like the drawing. Then the knot is tied ¾ inch
from the label, and the label tied to the right hind foot securely by a
square knot, and the loose ends trimmed off.

You must have a field catalogue. This is usually a looseleaf notebook containing the same information that you recorded on the label,
and in about the same order. It is really a duplicate label record
and should be guarded against loss. Should a label accidentally be
destroyed, it might be possible to identify it by this duplicate record.
You will also have a permanent record of the mammals you may
trade to other collectors.

The field notes are kept in another notebook. Here, individual needs can determine the information recorded. Some people make an entry for every specimen, recording additional information that is not on the label. Others make an entry for each trapping locality and night. In other words, if you set 29 traps for one night in a certain place, you record all the general data about that under a single entry in your field notes. You record weather, condition of the territory in which you are trapping, various habitats, kinds of bait, traps, how many specimens you caught, how many skinned, and the numbers of the skinned specimens, and always the number of trap-nights. A trap night is one trap set one night. So if you had 29 traps set for one night you had 29 trap-nights. If you set them in the same place for two nights, you had 58 trap-nights even though you had only 29 traps. You will think of other items you want in your notes. Do not make them too brief, for this is your record of activity in mammalogy. It is valuable, keep it always. If you ever give your collection to a museum, your notes go with the collection.

All entries on labels are to be made with black India ink, and done as neatly as possible. Entries in field catalogues and notebooks are best done in India ink as well, unless a typewriter is available.

SKINNING SMALL ANIMALS

It is probably best to use an animal the size of a chipmunk, squirrel, or wood rat for your first specimen, although when you gain some experience you will likely come to feel that meadow mice are the easiest of all, with pocket gophers running a close second.

Suppose that you now have a wood rat in your hand, and are ready to begin to make a study skin out of it. The first thing to do is to take the measurements in millimeters. The total length, length of tail, length of hind foot from the heel to the tip of the longest toe-nail, and the length of the ear from the notch are the usual ones to take (figs. 10 to 14). Bats have a fifth measurement, the length of the tragus,

Figure 10

Figure 11

Figure 12

Figure 13   Figure 14

a small rod-like structure in the ear. These are all recorded in order at the bottom of the front side of the label, thus: 290-130-29-25. The weight in grams is often added, and if one can afford a good balance, that is a worth while record to obtain. It is added after the measurements. If there is any chance that you might get the measurements mixed up it would be well to have them marked on the label. (See fig. 9).

The sex is now recorded, and the label completed before you start to skin the animal. Some mammalogists insist that the label be tied to the foot before skinning is begun. However, I find that the label often becomes soiled if it is attached to the specimen, but it may be laid to one side until the skin is ready to pin down, then it must be attached before you start the next specimen. This is important, for you must never get the label for one specimen attached to another specimen. There is no hurry about finding the name of the creature, so leave that until all the skinning is done. It is essential to finish the skinning as soon as possible, for it will be a better specimen if it is done quickly before it has a chance to spoil or to become dry.

Mammalogists differ in their preferences for skinning tools. Some use a scalpel or pocket knife almost entirely, while others use scissors. You might try both methods and choose the one you like best. I prefer scissors, so will describe the process with scissors.

Figure 15

Figure 16

With the scissors make an incision down the middle of the belly as shown in the drawing (fig. 15), then peel the skin back with the fingers until the knee joint is exposed. Cut through the knee by going through the joint rather than through the bone (fig. 16), then remove the flesh from the leg attached to the skin; do the same for the other leg, and pull the skin back toward the tail. During this time you will need to pour fine cornmeal or hardwood sawdust onto the skin to absorb moisture, to keep the fur clean, and to make it easy to get hold of the skin with your fingers.

Actually, most of the work is done by your fingers and you will appreciate the aid of the cornmeal.

Then, grasp the tail with the fingers and pull firmly (fig. 17). Be sure that you do not pull on the hair of the tail, but put the pressure on the tail bone below the skin, thus allowing the tail bone to pull out of the skin rather than pulling the skin off the tail bone.

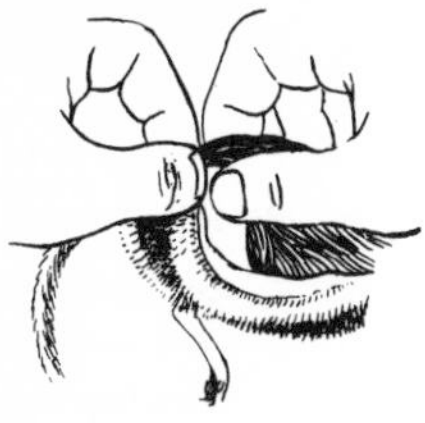

Figure 17

Now, it is a simple matter to peel the skin off the back of the animal until you reach the front legs. These will be severed at the joint as were the hind legs, and the peeling continued until the head is reached. It is necessary to be very careful in this peeling process so that you do not pull on the skin very much, for this may cause the skin to stretch and the animal to be much larger than it should when you are through. This is especially true of gophers and other thin-skinned animals.

Figure 18

Figure 19

Care must be used in cutting over the ears, for they could be cut off (fig. 18). The eyes are the next problem; if you hold the scissors very close to the bones of the skull and take very small cuts you will soon learn not to cut the eyelids (fig. 19). The teeth are a difficult matter in some animals, especially the rodents, for the mouth has two chambers, and care must be used not to cut through the skin. By keeping the scissors very close to the skull and cutting carefully you will soon have the skin at the very tip of the nose. Then, all you need do is to cut through next to the bone, and the skinning is done (fig. 19A).

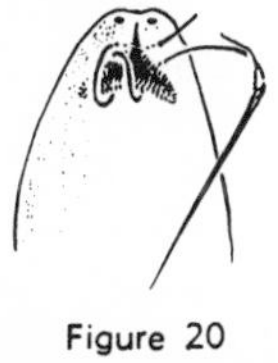

Next you will sew up the mouth. In rodents and rabbits the mouth opening forms the shape shown in the sketch.

Sew as the dotted lines indicate (fig. 20). Check over the skin now to make sure there are no bits of flesh clinging to it, and remove all fat present. Before going any farther dust a liberal supply of Boraxo all over the flesh side of the skin, before you place the cotton body in the skin and before you sew it up. This is to prevent insect damage later on.

Figure 19A

Figure 20

Figure 21

Take a wire the right size for the tail, wrap the base of it so as to fill out the tail, then insert it in the tail (fig. 21). Make sure that the wire goes clear to the tip, even if you have to use a file to make the wire small enough to go that far. If the wire does not reach the end of the tail, the tail will be broken off some time later when it is dry. Wrap enough cotton about the leg bones to make them the size of the original legs, and turn legs right side out. If a leg is broken, a wire, wrapped with cotton, should be placed in the leg to take the place of the bone.

Figure 22

Figure 23

You are now ready for the cotton body. Take a thin strip of cotton wide enough to extend from nose to base of tail, roll it up lightly, but rather firmly, and make a long slender roll a bit larger than the original body, and the same diameter throughout its length (fig. 22). Then with the tip of a pencil (the skin has been turned wrong side out all this time) poke a small dent in the nose, and, after pinching the cotton body down into a sharp point, stick this point into the dent made in the nose (fig. 23). Now turn the skin back slowly, making sure that the cotton is remaining in place, and filling out the head clear to the tip of the nose. If the skin has become a bit dry by this time, (as it may the first few times you do this) moisten it with water so that it will turn back freely.

Figure 24

Make sure that the cotton body completely fills the skin, and extends back as far as the base of the tail (fig 24). If it is too short, do not stick in small wads of cotton to fill in the gap, but remove the entire body and make another one that will be long enough. If the body is too long the excess part may be cut off with scissors, or torn off carefully. Pull the sides of the body over the cotton and bring them together, ready for sewing. Make sure that the tail wire fits well, and sticks outside the cotton on the belly side. See that the legs are in place. You are now ready to sew up the belly. If the skin does not come together moisten it with water so that it will cover the cotton easily.

Figure 25

Sew the belly skin together starting at the anterior end and making cross stitches about a half inch apart (fig. 25). A single thread is best, for it will not tangle so readily. When you are through just make a few extra stitches down through the skin, then cut off the thread near the skin. This should leave no visible thread. Now the specimen should be well stream-lined from tip of nose to end of body. Smooth out any bulges or dents, tie on the label, and pin down to a board or piece of cellotex as the sketch shows (fig 26). Leave there until dry.

Figure 26

The skull is removed from the body, and a small label, bearing your name, collector's number, and sex of the animal is attached carefully through the lower jaw (fig. 27). This is put away in a safe place to dry, protected from flies, as is the skin. The beginner may see no reason for so much attention accorded the skull. Determination of subspecies, and often species, many times depends on cranial differences.

Figure 27

If the skin is bloody or dirty it should be cleaned with a piece of cotton dipped in cold water. Dry the wet area by rubbing cornmeal over the fur until it is fluffy again. Never use warm water on blood.

Many mammalogists put powdered arsenic on the skin before the body is placed in position. Because of the danger of poison I do not recommend this practice for beginners. A live mammalogist is worth far more than a dead one! Boraxo is nearly as good as arsenic for preventing insect damage, and is never dangerous; use it instead of arsenic to protect your specimens from insect pests.

SKINNING LARGE MAMMALS

Mammals up to the size of woodchucks and badgers may be prepared the same as small mammals, with the added precaution that all the fat must be carefully removed, or the hair will come out of the skin. Larger animals such as coyotes, bob cats, deer, and the like are best tanned. The skinning can be done much the same as in the small mammals, but large skinning knives will be necessary, and the legs will be split down to the foot. The toes and claws must always be left on the skin. Animals larger than coyotes should have the skin on the underside cut all the way from the anus to the throat.

The skin should be tacked out on a large board, or against the side of a wall in the shade and left to dry. Flies should be kept away by spray.

As soon as possible the skin should be sent off for tanning. If the skin cannot dry quickly it would be wise to rub salt into the thicker portions around the lips, ears, feet, and any other area that does not dry readily. Measurements and data are kept in the manner previously described.

CLEANING SKULLS AND SKELETONS

After the skulls and skeletons are thoroughly dry, the next step is to place them in a suitable container with a colony of dermestid beetles. These are the little black beetles with silvery bellies that you may find under the dead carcass of an animal lying out on the ground. However, if that method of collecting the beetles does not appeal to you, you may be able to get your own colony by merely placing the

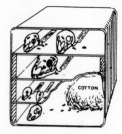

Figure 28

skulls or skeletons on a high window ledge or other place where they will not be in danger from cats or dogs, and where the beetles can find them. As soon as you have a few beetles present on the skulls place them all in a tight metal can of about five gallons capacity (fig. 28). Keep the can where it will remain at about room temperature for best results. Place a fairly large wad of cotton in the can, for the beetles will not pupate in the skulls. As soon as the new beetles hatch out they will be laying more eggs on the skulls, and then the colony will grow rapidly. In a week or so the skulls will be nicely cleaned and can be removed from the can. If you desire to maintain the colony permanently you will have to remember to keep them supplied with food. If you do not want to keep the colony then nothing more need be done until a new start is needed.

When the beetles have finished their work, the skulls may be placed in full strength commercial ammonia for a half hour or longer until the bones are nicely whitened. Then, wash them carefully in water and allow to dry before placing them in the permanent collection. The data label originally attached to the skull should be placed with the skull in a glass vial about the right size for the skull. Only one skull should be placed in a vial, so as to avoid any possibility of error. To make it doubly sure you may write the specimen number on the top of the cranium in India ink, but always keep the original skull label in the vial with the skull.

Large skulls like those of deer or sea lions may be cleaned by boiling and picking off the flesh. They may be whitened in ammonia like the others. Small skulls must not be boiled, for they may be seriously damaged.

If it is not convenient to make a beetle box for cleaning skulls or skeletons, they may be cleaned by another method. First place either freshly skinned or dried skulls in full strength commercial ammonia just as you buy it at the grocery store. Leave the skulls in this for several days, making sure that you have marked each skull so that you will not get them mixed up. It is best to place only one skull in a small bottle, with the specimen number marked on the outside of the bottle. After soaking in ammonia for several days, place the bottles with skulls in an electric cooker and heat to 150° for about 10 minutes for small skulls, and up to one hour for larger skulls. Small skulls cannot take more than a few minutes in the hot ammonia. Be careful not to breathe the ammonia fumes when you open the bottles. After the bottles have been in the hot water bath in the cooker for the required time, remove them from the heat, pour off the ammonia, rinse several times in clean water, then carefully rinse off and pick off all pieces of remaining flesh. Lay the skulls out to dry in a safe place. They will now be clean and white when dried. It takes a little experience to keep from injuring delicate skulls by this process.

PRESERVATION OF SPECIMENS

It is just as important to take good care of specimens as it is to prepare them carefully. First, make yourself a tight cabinet (fig. 29) with a removable door, and sliding trays which may be placed either close together or far apart as the case demands. The sides of the cabinet s h o u l d have a series of small cleats which allow the trays to be placed anywhere. Most museums make their cabinets of white cedar, and cover them with zinc. The tray size is usually about 24 x 41 inches, while the cabi-

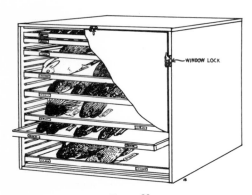

Figure 29

WINDOW LOCK

net is 42 inches high. Any other size or material that suits one's fancy may be used, of course. Be sure to build the cabinet in such a way that it will be absolutely pest proof, for dermestid beetles and clothes moths love to get into mammal collections. If the skins have been poisoned with arsenic less care will be needed, but even then one cannot be sure, so it is best to use plenty of napthalene flakes from time to time, or place cakes or large crystals of para-dichlorobenzine (PDB) in the trays. When the cabinet is to be closed for a long time it would be wise to place a fairly large quantity of PDB on the trays.

Arrangement is ordinarily made systematically according to order and families, placed in neat rows across the trays. Skulls should be placed next to the skins, or may be placed in cardboard trays near the skins.

LIFE HISTORY STUDIES

Along with the collection and preparation of specimens should go studies of various phases of mammalogy. A number of these have been mentioned above, such as home ranges, populations, and adaptations, but one of the most interesting and profitable studies with mammals is in regard to their life histories. The complete life histories of mammals are known in less than a third of the species, and in the western states the proportion is far less than a third. That implies that much work is yet to be done before we will know the details in the development of the young, in the time of breeding, the number of litters per year, food habits, and other phases of the life history of the mammal.

Yet it is not difficult for even a beginner to make observations along these lines. A few live traps will provide the necessary animals for study. Cages may be made from some hardware cloth and a wooden frame. Food may be easily provided in most cases, although the shrews and bats are not easy to feed since they are fond of living insects. Rodents are easy to raise, although even then there will be problems to overcome before one can have success with a group of wild animals in captivity.

For good examples of life histories I would refer the reader to the *Journal of Mammalogy*. This may be found in almost any large library. The twenty year index for the years 1919 to 1939 will be most convenient to find the exact references. A life history study contains data regarding the habitat where the animals live, the kinds of food they eat, the time of year they breed, descriptions of the nest, mating antics or habits, birth of the young, description of the young at all states of development with records of growth by daily studies of weight and measurements, details of development such as the time when the eyes are opened, when solid food is eaten, when teeth appear, when hair begins to appear on various parts of the body, when the young can walk or run about, and engage in other activities. In short, a life history study, as the name implies, records the details of the entire life of the animal from birth to death. It may be quite impossible for one to learn all these details from one family, in fact it would hardly be considered valid if only one family were studied, so it is usually necessary to raise several families of the same kind in order to complete the study.

This kind of study can be pursued in one's spare time. But while it takes only a little time each day, it is many days before the project is completed. Life history studies make fine thesis topics for candidates for master's and doctorate degrees, with the one disadvantage that the time when they may be completed is never certain. So many things can happen to spoil the study that one dares not predict whether or not he will ever finish the project. Since all this is true it is even more imperative that a number of students spend time on the life histories of our wild mammals. Each serious attempt adds something, at least, to our store of mammal knowledge. A study of the *Journal of Mammalogy* will generally reveal what has been done in the past, for this journal contains more material of this nature on mammals than any other source.

JOIN A MAMMAL SOCIETY

Everyone who is interested in mammals should become a member of the only mammal society in the world—the American Society of Mammalogists. Their journal (*Journal of Mammalogy*) is well worth the modest dues of $7.00 per year, for it contains usually more than 600 pages a year, and many of the articles are of interest to anyone —be he a professional mammalogist or not. New members are accepted upon recommendations by a member of the society.

PHOTOGRAPHY

 AMMAL photography is one of the most difficult forms of photography. One must combine a good deal of knowledge of photography and of mammalogy in order to succeed at this game. First a word about equipment, for you will need some, and the more the better, provided it is the right kind. You must decide first what kind of pictures you want, for that will be the chief thing to consider in choosing equipment. If you want to make large prints in black and white you will need a press type camera, a Speed Grafic or Graflex, or something similar with double extension bellows and flash attachment. Or, you may use one of the modern 2¼x2¼ cameras, with a variety of telephoto lenses and close-up focusing. This size can be used equally well for black and white prints or for color.

If you want motion pictures you will want a good quality camera taking either 8 mm., super 8 or 16 mm. film. In any case you will need

a camera that will take telephoto lenses, for seldom can you walk
up close enough to a mammal to get good pictures. You must nearly
always use a telephoto lens. For filming small mammals you will
often have to simulate the natural habitat of the mammal in your
own back yard by building artificial pens and cages which will
look like outdoor situations. Here you can use flood lights plugged
into your house in order to light up the artificial pen at night, since
nearly all such pictures will have to be taken at night.

For 35 mm. pictures in either black and white (for prints) or color
(for slides) you will need a flash unit on your camera unless you
decide to bring the animal to your home where you can use flood
lights. Even then, flash is usually better since it does not disturb
the animal, whereas it is difficult to get animals to behave normally
under strong flood lights. Most any brand of 35 mm. camera will be
satisfactory for these pictures, but often you will need a telephoto
lens, so buy only those cameras which will take telephoto lenses.
The telephoto lens has tremendous advantages for mammal photography, for it allows you to get a close-up picture of the animal without
your being close to the animal, therefore you do not disturb it, and
it reacts more normally. But telephoto lenses are expensive, and
so are the cameras that use them. For closeup mammal photography
it is usually best to use a single-lens reflex camera, of which there
are many on the market.

Flash pictures may be made with almost any camera, for the
flash units are versatile. Be sure to buy a good flash mechanism,
with the guarantee that the camera will be synchronized, for otherwise you may be wast-
ing your money for
flash bulbs and film
when the camera does
not work in perfect har-
mony with the flash
unit.   Flash pictures
are perhaps the most
interesting of all to
take. It is necessary
to use a certain amount
of ingenuity in order to
set up your camera so
that the animal will

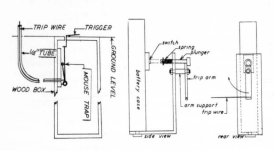

Figure 30

take its own picture. The sketches show (fig. 30) one way to build
a gadget that will trip the camera and flash unit, and even a shrew or
small mouse may set this off. Large animals will take their pictures
by merely walking into a fine wire or string placed across a trail,
and attached to the flash unit.

This mechanical release is designed for a flash unit with the
switch mounted on the battery case as shown in the drawing. By

making modifications in the bracket you may adapt this to almost any brand of flash unit. It would be best to put a shock-absorbing piece of sponge rubber or something else around the switch on the flash gun barrel, for such switches are seldom designed to stand hard shocks that they would receive from this trip mechanism. The spring should not be any stronger than necessary, for there is no need to place undue strain on the flash unit.

Many people prefer an electronic flash to the model using bulbs, but this is personal matter. I have used both, and have had far more trouble with electronic than with bulbs. If you buy an electronic flash, buy the best one available, and you are less likely to have trouble. The electronic flash will not be as convenient for an automatic setup where you let the animal take its own picture.

One should think of protecting the camera and equipment when taking night pictures of animals, for the hazards are many. Rain, wind, hail, snow, and even sand storms may come up while the camera is standing out in the weather. Large animals may do a lot of damage, too, especially if the equipment is supported by a tripod. It would be best to fasten the camera to a tree, or on a solid post in the ground. A covering could be placed over the camera to protect it from the weather.

Here are a few ideas for those who want to take close-ups of mammals in captivity. One of the simplest ways is to put the animal in a rather large pen, 6 or 8 feet in diameter, then get in the pen with it. The animal will crowd into one corner and will usually remain quiet. If you have strong flood lights you can take either color or black and white pictures. With a flash outfit one may dispense with the flood lights.

Another way to get close-ups is to build a glass cage, setting panes of plate or double strength glass into grooves in a 2 by 8 inch plank. The whole thing should be 3 feet or more in length so that you can get your pictures while the animal is in the middle of the cage. Natural conditions may be simulated by filling the cage with soil or sand and placing small plants in position inside, or else the cage may be set up in the natural habitat of the animal. Put screen over the top of the cage, for a glass cover will prevent evporation of moisture and will cause the glass to fog. You must have plenty of patience for making the exposures, for invariably the animal will crowd into one corner or the other—anywhere but in the middle. Yet it is only in the middle where you want to get your picture, for the corners of the cage will be sure to show otherwise. Here is where bright light and a fast shutter will come in handy, for you can get excellent action pictures as the animal races back and forth in the glass cage. If you can allow the creature to become calm, and place food in the middle of the cage, you may have better success. At least the method is worth trying.

SCIENTIFIC NAMES

I N the study of the small mammals it is not possible to find a group of standard common names such as we have for birds. It is therefore almost necessary to use the scientific names, which are not so bad when you get used to them. I would just as soon remember *Peromyscus* as white-footed mouse. *Microtus*, too, is certainly no harder than meadow mouse, after you learn it. In many cases the names are not so simple—*Clethrionomys*, for example —but in other cases they are very short, like *Zapus*, the jumping mouse. Some mammals have no common name at all, so there is no other choice in that case—*Phenacomys* is an example.

The names are easy to pronounce once you get used to them. Remember that the accent is usually on the syllable third from the end; thus *Phenacomys* is Phen-a-co-mys (fen-á-co-miss). The rule does not always hold, however, for *Microtus* is Mi-cró-tus, and *Peromyscus* is Per-o-mýs-cus. The syllable "mys" which often occurs in these names means mouse. Remember that not everyone agrees on the proper way to pronounce many of the names, so do your best, and let it go at that unless you find you have been wrong.

Scientific names of mammals are like those of other animals. First comes the genus name, which is always spelled with a capital, then the species follows, and the subspecies. These last two names are never spelled with capital letters.

The three together make up the scientific name of the animal. Besides these, however, are the family names and order names, for all mammals belong to the class Mammalia of the Vertebrate subphylum, and the class is further subdivided into orders. A key to these orders is the first key in the book. Each order is in turn divided into families, and the families into genera (singular, genus). The genera contain species, and the species are made up of subspecies.

A number of technical terms are needed in the keys, but these have been kept to a minimum. The glossary-index will explain most of these; for any others, try the dictionary!

HOW TO USE THE KEYS

 KEY is a short-cut to the name of the mammal you are working on. Do not try to find your mammals by looking first at the pictures. Your quickest and most reliable method is to use the keys. First, a key consists of a series of opposite statements beginning with very general characteristics, and finally working down to the more specific differences of the mammals. If you have the specimen before you, you will have no trouble working the keys. This will usually be the case, for it is seldom that you can go out into the woods or fields and key a mammal from a distance. This is where mammalogy becomes easier than ornithology, for you generally have the specimen in your hand.

You will begin with the key to the orders if you do not have the slightest notion what your specimen is. However, as soon as your knowledge increases a bit you will be able to tell the order without resorting to the keys. Then you will turn at once to the order in question and begin with the family key. Eventually you will also know the families by sight, and can turn at once to the key to the genera under the correct family. The keys in this book are divided into these groups so that it will be easier for you to find your animal. In other words, when you know the orders and families you will no longer need to begin at the very first key.

You will begin with numbers 1a and 1b of the key to the orders; by reading both statements you will decide that only one of them fits your specimen, and the correct statement will refer you (by number) to the next series of statements you must read, usually another page farther along in the book. Here you may find a description of the order, and a key to the families. You will begin at 1a and 1b again, choose the correct statement, and proceed as before to the family. Again you will turn to another page and find the genus and the species in the same manner. At the species write-up you will find a more complete description along with the small outline map showing where the species lives in the United States. For identification to subspecies, and for further information about the animal, you will have to refer to the books and references listed on pages 11 and 12.

Let us try an example. You have been trapping high in the rock slides of the Rockies of northern Montana, and have taken a small animal with short ears, and tiny feet, but without a sign of a tail. Begin at the Key to the Orders, page 39 with the first pair of statements. Since the forelimbs are not wings, you go to number 2. You

discover that 2b fits this animal, and refers you to 4. As the animal
has no hoofs, you choose 4b, which sends you on to 6. In turn, you
choose 6a, 7b, 8a, 9a, and, with a good look at its incisors, you arrive
at 10a—the *Lagomorpha,* the order of the rabbits and pikas, or conies.

Now you go to page 69 and read the description of the *Lago-
morpha.* Keying further, you discover that since the specimen has
no tail, it belongs to the *Ochotonidae.* There follows a description of
this family, and since there is only one genus and one species, you
have now found the name and correct description of your pika,
Ochotona princeps.

You will occasionally find keys referring to characteristics of
skull bones and teeth. I have tried to use other characters whenever
possible, but at times even genera overlap so much in general characteristics that the more technical differences have to be used in
the keys. It is for this reason that it is so necessary to save the
skull for every specimen. In a few cases I have been able to save
only a single jaw bone from some of my specimens which have
been badly damaged, yet this is sometimes enough to help identify
the creature correctly.

A beginner in the reading of mammal keys may be a bit per-
plexed, and not a little amused as he notes that the back of a rabbit
may be "washed with pinkish cinnamon," or that a mouse may be
ochraceous tawny in color. Such shades of color have not been
plucked from the writer's fevered imagination. In 1912 a man by
the name of Ridgway worked out a color chart showing all possible
shades that an animal might possess. He named these hundreds of
different colors as well, and it is to this *Color Standards and Color
Nomenclature* that a mammalogist comes when he describes a mammal for the first time. After this, every other writer uses this original
description when writing about the mammal.

KEY TO THE ORDERS OF NORTH AMERICAN MAMMALS
(Including Common Domestic and Zoo Animals)

1a. Forelimbs modified as wings, Fig. 31. CHIROPTERA, bats, page 57.

Figure 31

1b. Forelimbs not modified as wings.............................2

2a. Hind limbs absent 3

2b. Hind limbs present 4

3a. Form fish-like, with the tail flattened horizontally, notched in the middle, and developed into "flukes"; eyes placed laterally on the head without binocular vision. Fig. 32. CETACEA, whales, porpoises, and dolphins, page 180.

Figure 32

3b. Form not fish-like, but resembling a sea lion to some extent; tail rounded and flattened horizontally, without a notch or "flukes"; eyes in the front of head, with binocular vision. Fig. 33. SIRENIA, sea cows or manatees, page 191.

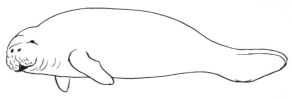

Figure 33

4a. Feet provided with hoofs...................................5
4b. Feet provided with nails or claws.........................6
5a. Hoofs divided into pairs, two large toes and two small ones ordinarily known as "dew claws"; cattle and deer-like animals. Fig. 34. ARTIODACTYLA, deer, cattle, sheep, goats, pigs, page 191.

Figure 34

5b. Hoofs single, as in the horse, or else toes uneven in number, as in the tapir, or rhinoceros. Fig. 35. PERISSODACTYLA, page 191.

Figure 35

6a. Feet provided with claws...................................7
6b. Feet provided with either claws or nails, but if claws, then they are not typical, but are nail-like, resembling human nails to some extent; fore feet hand-like with opposable thumb; hind feet developed into a rather large flat sole, similar to that of man; eyes large, and placed in large orbits; vision typically binocular, like that of man. Fig. 36. PRIMATA, man, apes, monkeys, lemurs, page 191.

Figure 36

7a. Body covered with bony plates forming a basket-like shell. **Fig. 37.** EDENTATA or XENARTHRA, the armadillo, page 179.

Figure 37

7b. Body not covered with bony plates..........................8

8a. Teeth present 9

8b. Teeth absent; mouth beak-like, even duck-like; body covered with dense fur with the feet webbed, or body covered with spines; these are the only mammals that lay eggs. Found in Australia and Tasmania, and in some of the larger zoos. **Fig. 38.** MONOTRE-MATA, duck-billed platypus and spiny anteater.

Figure 38

9a. Canine teeth absent; incisors chisel-like....................10

9b. Canine teeth present; incisors not chisel-like.................11

10a. Two pairs of incisors present in upper jaw, second pair much reduced and located directly behind first pair. **Fig. 39.** LAGO-MORPHA, rabbits, hares, and pikas, page 69.

Figure 39

**10b. Only one pair of incisors in upper jaw. Fig. 40. RODENTIA,
mice, rats, squirrels, chipmunks, woodchucks, beavers, page 77.**

Figure 40

**11a. Canine teeth longer than other teeth, sometimes tusk-like; eyes
well developed** 12

**11b. Canine teeth similar to other teeth; eyes inconspicuous or appar-
ently absent. Fig. 41. INSECTIVORA, shrews and moles, page 44.**

Figure 41

**12a. Tail prehensile; first digit on fore and hind limbs opposable, or
thumb-like; marsupium, or brood pouch, present on underside of
abdomen in females. Fig. 42. MARSUPIALIA, opossums, kanga-
roos, koala bears, page 43.**

Figure 42

12b. Tail not prehensile; first digit not opposable; no marsupium...13

13a. Hind feet developed into flippers; habits mainly aquatic, except in breeding on islands. Fig. 43. PINNEPEDIA, seals, sea lions, and walruses, page 174.

Figure 43

13b. Hind feet not developed as flippers. Fig. 44. CARNIVORA, dogs, foxes, cats, weasels, bears, raccoons, page 157.

Figure 44

Order MARSUPIALIA

Opossums

This order includes mammals which bring forth their young in a very premature state, which are then attached to the mammae in the marsupium until they are more fully developed. In the United States there is only one native marsupial, the opossum, but in Australia a large group of marsupials occur. The kangaroo, wallaby, koala bear, and flying phalanger are just a few, and several of these may be seen in zoos. In Central and South America are several more types like the water opossum and the coenolestid, both of which resemble our opposum to some extent.

Fig. 45. *Didelphis marsupialis.* Opossum.

Figure 45

Total length 780 mm.; tail 300 mm.; hind foot 70 mm.; weight 6 to 8 pounds. Tail prehensile, or with the ability to curl around objects; ears round and naked; first digit opposable; females with a marsupium or brood pouch in which the young remain for a number of weeks after birth (a dozen or more new-born opossums can be held in a teaspoon!); fur long and lax, gray or blackish in color with creamy underfur.

Opossums will sometimes play dead if they are handled, but more often they will show their teeth, hiss, and give the appearance of being very ferocious. This seems to be all bluff, for the animal will seldom attempt to bite; a look at its teeth will cause one to respect that part of the animal, in any case. While opossums prefer flesh food, they will eat anything in sight. They are lazy creatures, preferring to inhabit an old squirrel's nest or a skunk's den rather than to prepare a nest themselves.

Order INSECTIVORA

Shrews and Moles

The insectivores are flesh-eating mammals of small size, with fine thick fur, long slender noses, inconspicuous ears, inconspicuous eyes and teeth which are nearly all much alike in that they are sharp and needle-like, and much the same length. The molars are somewhat wider than the other teeth. The skull is elongated with the brain case short and the zygomatic arch slender or incomplete. At least one member of this order can be found in nearly every part of the United States.

Key to the Families of INSECTIVORA

1a. Forefeet wide and spade-like, developed into palms for digging; ears inconspicuous; eyes apparently absent, but present under the skin; zygomatic arches complete. *Talpidae,* moles, page 45.

1b. Forefeet slender, not developed for digging; ears somewhat conspicuous; eyes present but small and inconspicuous; zygomatic arches incomplete. *Soricidae,* shrews, page 48.

Family TALPIDAE
Moles

Moles occur in nearly all parts of the United States, burrowing in the ground in somewhat the same manner as do the pocket gophers. Moles push up the dirt in a ring around the hole while gophers usually push up dirt at one side of the hole only. This characteristic is not always plain enough to distinguish, however.

The body is stout and cylindrical, covered with a thick coat of fine soft fur; external ears are absent; eyes small and either well concealed in the fur, or sometimes not even appearing through the skin, but always present on the skull; snout proboscis-like and flexible; fore feet spade-like, hind feet not spade-like. Nests are usually placed from 12 to 19 inches below the ground, in a maze of deep tunnels. They also make the familiar surface tunnels which just push up a ridge of dirt under the surface of the ground. This surface tunnel is made in search of food—earthworms and insect grubs. Some of the diet is formed of tender plant shoots, but most of the damage done to plants is actually done by gophers, not moles.

Key to the Genera and Species of Moles

1a. Found only west of the Rocky Mountains......................2

1b. Found only east of the Rocky Mountains......................5

2a. Length of tail less than one-fourth the total length; width of palm of fore foot equaling or exceeding length of palm; total length more than 150 mm.; teeth, 44. *Scapanus*...........................3

2b. Length of tail more than one-fourth of total length; width of palm less than length of palm; total length 125 mm. or less; teeth 36. Fig. 46. *Neurotrichus gibbsii*. Shrew-Mole.

Total length, 110 to 125 mm.; tail 40 to 45 mm.; hind foot 17 mm. Color sooty black or sooty brown, slightly paler on the underparts; fur long but not very thick.

Figure 46

Shrew-moles are shrew-like animals found most commonly in swampy places, and in damp woods. They may be caught in traps set about rotting logs, or their small tunnels may be found among the rotting leaves in marshy places.

Figure 47

3a. Size about 200 mm.; greatest length of skull more than 40 mm. Fig. 47. *Scapanus townsendii*. Oregon Mole, Townsend Mole.

Total length 200 to 235 mm.; tail 40 to 50 mm.; hind foot 27 mm. Color blackish brown to sooty black with a purplish sheen; underparts slightly paler; skull large with heavy mastoids and long rostrum.

This mole is the largest in the United States, and makes extensive burrow systems in the fields of the coastal areas of the Pacific states. Its fur is of good enough quality to be valuable commercially, and its size is large enough to make it worth catching for fur when prices are high.

3b. Size less than 200 mm., often much less than 200 mm.; length of skull less than 40 mm. 4

4a. Total length 170 mm. or less; tail 30 to 50 mm.; color dark sooty black to sooty brown; feet and claws smaller than 4b; tail and feet less hairy than 4b. Fig. 48. *Scapanus orarius*. Coast Mole.

Figure 48

The habits of all the western moles are much alike. There are three subspecies of this mole in Washington, Oregon, and northern California.

Figure 49

4b. Total length 170 to 190 mm.; tail 30 to 40 mm.; color pale silvery drab or even brassy brown, the underparts still paler. Some of the subspecies are even paler than this. Fig. 49. *Scapanus latimanus*. California Mole.

5a. Snout with 22 fleshy, finger-like tentacles; tail long, 60 mm. or more in length; skull 33 to 35 mm. long, slender and weak; eyelids do not cover eyes permanently. Fig. 50. *Condylura cristata*. Star-nosed Mole.

Total length 170 to 200 mm.; tail, 60 to 85 mm.; hind foot 25 to 29 mm.; teeth, 44; color either black or very dark brown.

This odd mole gets its name from the fleshy finger-like tentacles at the tip of the snout. The front feet are wide as usual, but are not so large as are those of the Prairie Mole. The long tail is constricted at the base, and is enlarged toward the middle by a mass of stored fat. This mole comes out of the ground more frequently than other moles, even enjoying a swim in a pond.

Figure 50

5b. Snout plain, no tentacles present; tail less than 45 mm. in length..6

6a. Tail naked; front feet broader than long; fur finer and softer than in *Parascalops*; teeth, 36. Fig. 51. *Scalopus aquaticus*. Prairie Mole.

Total length 150 to 190 mm.; tail, 21 to 38 mm.; hind foot, 18 to 20 mm.; weight 66 to 118 grams. Color blackish brown with a silvery sheen, paler in summer; underparts grayish.

This mole almost never comes to the surface of the ground, except to push dirt out of its burrow; it makes tunnels at two levels, one just beneath the surface, which is only a ridge pushed up as the animal searches for worms and insects, while the other tunnels are 12 to 24 inches below the surface. Only about 15 percent of the diet of this mole is vegetable matter, so it is not nearly so harmful as some may think.

Figure 51

6b. Tail densely furred; front feet about the same width as they are in length; fur coarser than in *Scalopus;* teeth 44. Fig. 52. *Parascalops breweri.* Hairy-tailed Mole; Brewer Mole.
Total length 150 to 170 mm.; tail 24 to 30 mm.; hind foot 17 to 21 mm.; weight 40 to 64 grams. Color slaty black, paler on the underparts.

Figure 52

This mole is similar in habits to the others, but it does not remain in its burrows so constantly as does the Prairie Mole, for it seems to enjoy coming out at night and wandering about in the forests in search of insects. Actually, the animal consumes large quantities of insects, and thus deserves some credit for this habit. It does very little damage since it occurs mainly in mountainous areas, but lawns or golf greens are sometimes harmed by its tunnels.

Family SORICIDAE

Shrews

Shrews are much smaller than moles, with the one exception of the shrew-mole, *Neurotrichus gibbsii.* The body of a shrew is more slender than that of a mole, the fore feet are not adapted for digging in particular, and the eyes are more prominent; external ears are present; zygomatic arch of skull absent. Four genera are recognized at the present time, with more than 30 species.

Key to the Genera and Species of Shrews

1a. Tail short, less than 30 mm., and less than half the length of head and body; ears either concealed or visible....................2
1b. Tail long, more than 30, and longer than half the length of head and body; ears plainly visible...................................4
2a. Ears concealed in the fur; teeth 30 or 32......................3
2b. Ears plainly visible; teeth 28, only 3 unicuspids visible. Fig. 53. *Notiosorex crawfordi.* Crawford Shrew.

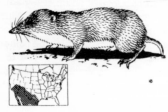

Figure 53

This shrew inhabits dry places where there is almost never any trace of water, often living in the sagebrush or in the creosote bush covered hills of the Southwest. Very little is known about its habits or life history. It is one of the rarest of mammals.

**3a. Length less than 90 mm.; color brown; teeth 30; only
3 unicuspids visible from the side. Figs. 54, 55,**

Cryptotis parva. **Least Shrew.**
Total length, 69 to 84 mm.; tail 13 to 17 mm.; hind
foot, 9 to 11 mm.; weight 4 to 6½ grams; color almost uniformly dark brown with ashy gray underparts.

UNICUSPIDS

Figure 54

Figure 55

These little shrews ordinarily live
in open regions where there are no
logs or stumps, where they probably
nest in holes in the ground much
like the Prairie Mole. In fact, they
may be found in the runways of
the mole. Little is known about
the habits of this shrew, but it probably feeds on insects and worms like
the other shrews.

Another species, *Cryptotis floridana,* occurs in Florida. It is
slightly darker in color, more grayish, and has the tail about 22 mm.
long.

**3b. Length more than 90 mm.; color slaty black; teeth
32; four unicuspids visible from the side, the fifth
very small. Figs. 56, 57.** *Blarina brevicauda.* **Short-
tailed Shrew.**

Total length, 98 to 132 mm.; tail 17 to 30 mm.; hind
foot 13 to 17 mm.; weight 12 to 23 grams; length of
skull 21 to 24.8 mm.; color dark slate gray; body
stocky.

UNICUSPIDS

Figure 56

Figure 57

This shrew inhabits the forests
where it prefers low swampy places,
but it may occur almost anywhere in
a forested region. They make long
series of burrows just under the surface where the decaying leaves thick-
ly cover the ground. Here they feed
upon insects in great numbers. There
are about 8 subspecies, and a closely
related species, *Blarina telmalestes,*
occurs in southern Florida.

4a. Total length 80 to 85 mm.; unicuspids 5, but in superficial lateral view appearing to be only 3, the third and fifth being almost invisible. Figs. 58, 59. *Microsorex hoyi*. Pigmy Shrew.

AmericAN?

UNICUSPIDS

Figure 58

Total length 78 to 103 mm., but usually nearer 85 mm.; tail 30 to 34 mm.; hind foot 8½ to 11 mm.; weight 2½ to 4 grams; length of skull 15 to 16½ mm. Color light brown with paler underparts.

Figure 59

This miniature shrew cannot be distinguished safely from the other small long-tailed shrews without noting the teeth. It is found in wooded areas usually near water or marshy places. Almost nothing is known about the habits or the life history of this tiny shrew. In most places it is a very rare animal, and the capture of a specimen is worth a celebration.

4b. Total length 90 mm. or more; unicuspids 5, and appearing to be 5 since the third and fifth are plainly visible from the lateral view. Fig. 60. Sorex. Long-tailed Shrews 5

UNICUSPIDS

Figure 60

5a. Hind foot 18 mm. or more; total length usually about 150 mm.; tail about half of total length; color grayish, never distinctly brown 6

5b. Hind foot less than 18 mm.; total length usually about 100 mm.; tail varies, but never as long as 75 mm.; if hind foot is over 16 mm. then the color is distinctly brown.............................7

6a. Distinct fringe of hairs along margin of hind foot enabling the creature to swim; rostrum only slightly down-curved; anterior end of premaxilla scarcely narrower dorso-ventrally than middle portion; size somewhat smaller than 6b. Fig. 61. *Sorex palustris.*

Northern

Water Shrew or Navigator Shrew.

Total length 142 to 160 mm.; tail 63 to 76 mm.; hind foot 19 to 21 mm. Color brownish black with silvery underparts; tail bicolor, or dark above and silvery below.

The water shrew inhabits mountainous or wooded areas, and stays along streams, lakes or ponds, or even in marshy places. If it decides to go to the other side of the stream or pond, it merely swims across, for the fringed hind feet make good paddles. The food is made up of a variety of insects, snails, and other aquatic and semi-aquatic animals. Some have actually seen these animals run across the water for a distance of five feet without breaking through the surface tension of the water, in

Figure 61

much the same fashion as a water skipper or spider. Very little is known of the life history of the water shrew.

6b. Fringe of hairs along margin of hind foot not very distinct; rostrum distinctly down-curved; anterior end of premaxilla much narrower dorso-ventrally than middle portion; size somewhat larger than in

PACIFIC WATER SHREW

6a. Fig. 62. *Sorex bendirii.* Bendire Shrew.

Total length 155 to 170 mm.; tail 61 to 80 mm.; hind foot 18½ to 21 mm. Color almost black in winter, but very dark blackish brown in summer; underparts usually about the same color as upper parts—lacking the silvery color of the Water Shrew, but in one subspecies *(Sorex bendirii albiventer)* the underparts are whitish as they are in the Water Shrew *(Sorex palustris).*

Figure 62

Little is known of the habits of this shrew, but its frequents wet places as does the Water Shrew. It also swims readily, but since the fringe of hairs around the hind feet are not pronounced it probably cannot swim as well as does the Water Shrew.

7a. Third unicuspid smaller than the fourth. Fig. 63   8

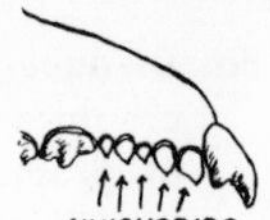

UNICUSPIDS

Figure 63

7b. Third unicuspid not smaller than the fourth. Fig. 64 ....**14**

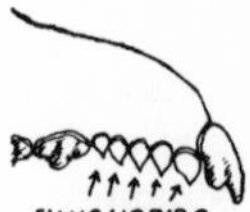

UNICUSPIDS

Figure 64

8a. Geographic range east of the Mississippi River. Fig. 65. *Sorex longirostris*. Bachman Shrew.

Figure 65

Total length 79 to 90 mm.; tail 27 to 30 mm.; hind foot 10 to 10½ mm. Color brown; rostrum short with the unicuspid row crowded; first and second unicuspids large, third and fourth smaller, fifth very small.

This shrew is very rare. Evidently it prefers damp places, but it has also been found in drier situations. Hamilton says to set traps in clumps of honeysuckle in the winter.

8b. Geographic range west of the Mississippi River **9**

9a. Underparts scarcely, if any, paler than the upper parts; tail sharply bicolor. Fig. 66. *Sorex trowbridgii*. Trowbridge Shrew.

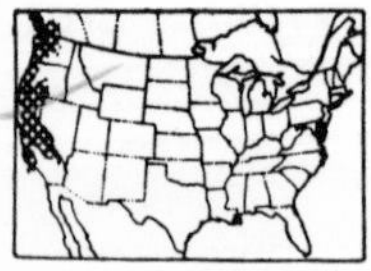

Figure 66

Total length 113 to 133 mm.; tail 52 to 61 mm.; hind foot 13 to 15 mm. Color dark mouse gray on both back and underparts; tail dark above, almost white below.

This shrew lives in damp places around logs and old stumps. Traps set in these places, baited with rolled oats, are effective in capturing many specimens where this shrew is common.

9b. Underparts of body distinctly paler than upper parts; tail not sharply bicolored 10

10a. Foramen magnum placed relatively ventral, encroaching less into supraoccipital and more into basioccipital. Fig. 67...........................12

Figure 67

10b. Foramen magnum placed relatively dorsad, encroaching more into supraoccipital and less into basioccipital. Fig. 68...........................11

Figure 68

11a. Cranial breadth less than 7.4 mm. Fig. 69. *Sorex tenellus.* Dwarf Shrew.

Total length 85 to 95 mm.; tail 36 to 41 mm.; hind foot 10 to 12½ mm. Color drab with underparts pale smoke gray; tail olive brown above, slightly paler below.

Figure 69

This shrew was known for many years from only one specimen, but mammalogists from the University of California discovered a number of them in western Nevada, where they lived in damp shady places near a small stream. One specimen was caught on a south exposure among boulders where it was very dry, and the only plants were sagebrush, rabbit brush, and Ephedra.

Sorex myops (White Mountains, California), and *Sorex nanus* (mountains of central and northern Colorado) are so similar to *Sorex tenellus,* that they will not be described here.

11b. Cranial breadth more than 7.4 mm. Fig. 70. *Sorex ornatus.* California Long-tailed Shrew.

Figure 70

Total length 89 to 108 mm.; tail 32 to 43 mm.; hind foot 12 to 13 mm. Color brown; underparts smoke gray; tail indistinctly bicolor, dark brown above, lighter below.

The habits of this shrew are much like those of others which live in damp woody places.

12a. Total length 140 to 150 mm.; underparts nearly as dark as the back; skull large and broad with heavy rostrum; teeth heavy with unicuspids broad and swollen. Fig. 71. *Sorex pacificus.* Pacific Shrew.

Tail 59 to 64 mm.; hind foot 16 to 17½ mm. Color brown to brownish black, underparts slightly paler; tail all one color.

Figure 71

These large reddish brown shrews are found in marshy areas where there are many fallen logs; they also occur in the deep forests where it is not so damp. Their food consists largely of insects.

12b. Total length from 100 to 135 mm.; underparts distinctly paler than the back; skull smaller and narrower with weaker rostrum than in 12a; teeth smaller and lighter.................13

13a. Total length usually more than 110 mm.; tail usually more than 44 mm.; least interorbital breadth usually more than 3.3 mm.; brain case raised posteriorly. Fig. 72. *Sorex obscurus.* Dusky Shrew.

Figure 72

Total length 103 to 135 mm.; tail 41 to 62 mm.; hind foot 13 to 15 mm. Color brown to brownish drab; underparts distinctly paler, pale smoke gray; tail bicolor, brown above, nearly buff below.

The dusky shrew lives along streams in the mountains, and in wet mossy meadows, or in places where spring water seeps out over rocks forming masses of moss, liverworts, and a host of water-loving plants. Here the shrews make short tunnels among the matted vegetation, winding in and out among the dripping plants. Traps set along these streams in mountain meadows are almost sure to reveal the Dusky Shrew.

13b. Total length usually less than 110 mm.; tail usually less than 44 mm.; least interorbital breadth usually less than 3.3 mm.; brain case flattened posteriorly. Fig. 73. *Sorex vagrans.* Vagrant Shrew.

Total length 96 to 110 mm.; tail 38 to 44 mm.; hind foot 11 to 13 mm. Color brownish black to warm brown; underparts pale smoke gray; tail scarcely bicolor.

Figure 73

This is the most common shrew in the western states, and occurs almost everywhere that water is found. It frequents small streams even in desert places, but is most common in marshy regions, in forests along streams, and around margins of lakes. Insects make up the diet of this shrew. Several subspecies occur, those of the coastal area being much darker than those of inland arid regions.

14a. Infraorbital foramen with posterior border lying caudad to plane of interspace between first and second molars. Fig. 74. *Sorex dispar.* Gray Long-tailed Shrew.

Total length 121 to 130 mm. tail 55 to 60 mm.; hind foot 14 to 15 mm. Color dark mouse gray, just slightly paler on the underparts; tail dark above, and only slightly paler below.

Figure 74

This shrew might be confused with the Smoky Shrew (*Sorex fumeus*) but has a longer tail, and is almost uniform color all over. It inhabits coniferous forests where it prefers damp places among boulders and damp mossy logs. Almost nothing is known of its habits.

14b. Infraorbital foramen with posterior border lying even with or anterior to plane of interspace between first and second molars **15**

15a Maxillary breadth less than 4.6 mm. **16**

15b. Maxillary breadth more than 4.6 mm. **17**

16a. Condylobasal length 15 mm. or more. Fig. 75. *Sorex cinereus.* Cinereus Shrew.

Total length 85 to 111 mm.; tail 30 to 48 mm.; hind foot 11 to 14 mm. Color grayish brown with the underparts light smoke gray; tail bicolor, brown above and buff below.

This little shrew is so versatile in his choice of habitats that he may be found from the salt marshes along the sea, through the wet meadows and marshy places in the valleys, in the damp woods, and clear to the highest mountain meadows. Insects make up his diet.

Figure 75

16b. Condylobasal length less than 15 mm. Fig. 76. *Sorex preblei.* Preble Shrew.

Total length about 95 mm.; tail about 36 mm.; hind foot 11 mm. Color brownish gray; underparts pale smoke gray; tail dark above, somewhat lighter below.

Since only three specimens have been found altogether, there is very little known about this shrew. It seems rather sure that they live in marshy areas in the yellow pine forests of eastern Oregon. This is one of the rarest animals in the world.

Figure 76

17a. Condylobasal length more than 17.5 mm.; cranial breadth 8.5 mm. or more; maxillary tooth row 6.1 mm. or more...........18

17b. Condylobasal length less than 17.5 mm.; cranial breadth less than 8.5 mm.; maxillary tooth row less than 6.1 mm. Fig. 77. *Sorex merriami.* Merriam Shrew.

Total length about 90 mm.; tail about 35 mm.; hind foot about 11½ mm. Color drab, paler on the flanks, with whitish underparts; tail bicolor, brown above, whitish below.

Only a few of these rare shrews have ever been found, and most of them were caught by the can or bottle method (see page 15). This is one of the few shrews which lives in dry areas. It may be much more common than has been suspected, for George Hudson of Washington State University, has caught a good many of them recently.

Figure 77

18a. Tail less than 45 mm.; coloration distinctly tricolor, for the back is much darker than the sides, while underparts are still much paler. Fig. 78. *Sorex arcticus.* Saddle-backed Shrew.

Total length 108 to 117 mm.; tail 38 to 42 mm.; hind foot 14 mm. Color blackish brown on the back, sides snuff brown contrasting sharply with the color of the back; underparts paler and more grayish than the sides.

Figure 78

Nothing is known of the habits of this shrew, except that it inhabits marshy places as do most of its relatives. Records of from six to nine embryos have been noted from specimens caught in April and early June.

18b. Tail more than 45 mm.; color only bicolor—back and sides the same color; underparts paler. Fig. 79. *Sorex fumeus.* Smoky Shrew.

Total length 110 to 127 mm.; tail 37 to 47 mm.; hind foot 12 to 14 mm. Color dark gray with underparts paler gray; tail indistinctly bicolor, darker above, paler below.

Figure 79

This shrew prefers the damp woods where logs are covered with moss, where clumps of yew or masses of ferns conceal wet springy places. It travels the tunnels of *Blarina* (the Short-tailed Shrew) or of red-backed mice. It is not found in the salt marshes along the coast, nor does it occur in dry areas. Sometimes it is common in a locality, then again no specimens may be found for a long time in the same area.

The following additional shrews occur in the United States, but are not shown in the keys. Consult Hall and Kelson (1959) for information on these species.

Sorex lyelli	*Sorex willeti*
Sorex trigonirostris	*Sorex sinuosus*
Sorex nanus	*Sorex alascanus*

Order CHIROPTERA
Bats

Bats are among the most feared, yet the least known of American mammals, as far as the average person is concerned. Some still have the notion that bats get into one's hair, carry bedbugs, or are omens of some evil that may come to them. Actually, bats are among the most valuable creatures we may have about our homes. Since nearly all American bats feed on insects, and since it takes so many insects to keep a bat alive, it follows that these animals are worth cultivating about our homes.

The wings of bats are actually only thin membranes stretched between the elongated fingers of the forelimbs. The wing is continuous with another membrane, the interfemoral membrane, between the hind limbs, which encloses the tail. A stiff spur, the *calcar*, extends from the heel along the margin of the interfemoral membrane.

Ears of bats are usually very large, for bats get around in the dark by use of an ingenious form of radar. As they fly, they produce sounds out of our range of hearing, and the sound waves from these notes are reflected back by any object they strike. The big ears of the bats catch these vibrations and the bat is warned of the obstacle. Bat eyes are so small that they must be useful only in detecting the approach of daylight, so that the bat knows when it is time to head back to its cave. Bats which have been banded during their day-time sleep in Carlsbad Caverns (New Mexico) have been taken 150 miles away from the Caverns at night. Estimate the tons of mosquitoes consumed over a radius of 150 miles about Carlsbad by the millions of bats which regularly inhabit the Caverns.

Bats may be found in attics of old houses, in out buildings, in caves, clinging to the bark of trees or hanging from limbs, or under large boulders. Not a great deal is known about their life histories, for they are difficult to find, hard to collect, and hard to keep in captivity.

Key to the Families of Bats in the United States

1a. Third finger of wing with 3 phalanges; leaf-like projection on nose in most genera; found only in the southern part of the United States. *Phyllostomidae*, page 58.

1b. Third finger with only 2 phalanges; no leaf-like projection on nose; found either in the south or throughout the U. S. 2

2a. Ear with a tragus; tail not extending beyond membrane; found throughout the U. S. and Canada. *Vespertillionidae*, page 61.

2b. Ear without a tragus; tail extending some distance beyond interfemoral membrane; found only in the extreme southwest. *Molossidae*, page 60.

Family   PHYLLOSTOMIDAE

This is the vampire family, although none of those found in the United States have blood-sucking habits. The nose usually has a leaf-like projection; tragus present; tail very long; wing membrane reaches to the ankle.

It may be necessary to use a magnifying lens to see the teeth characteristics which are used in the following key. External characters are not sufficiently marked to differentiate between genera.

Key to Genera and Species of *Phyllostomidae*

Figure 80

1a. Nose-leaf absent. *Mormoops.* With further characters as follows. Fig. 80. *Mormoops megalophylla.*
Total length 90 mm.; tail 28 mm.; forearm 56 mm.; color brown; nose-leaf absent, but dermal outgrowths on chin very much pronounced; skull so greatly shortened that both rostrum and braincase are broader than long; braincase so greatly deepened that its floor is elevated until the lower rim of the foramen magnum is above the level of the rostrum.

1b. Nose-leaf present 2

2a. Upper molars normal, that is the "W" pattern of the cusps and ridges is evident (Fig. 81). *Macrotis.* With further characters as below. Fig. 82. *Macrotis californicus.* California Leaf-nosed Bat.

Figure 81

Figure 82

Total length 93 to 103 mm.; tail 33 to 41 mm.; hind foot 14 to 17 mm.; forearm 46.8 to 52.6 mm.; color pale brownish gray; ear so long that it reaches beyond tip of muzzle when laid forward; long leaf-like appendage on nose; tragus short and slender.
This bat inhabits the low hot deserts.

It rests in caves during the day; at night it captures beetles, flies, and other flying insects.

2b. Upper molars abnormal, that is the "W" pattern of cusps and ridges is either absent or not evident..........................3

3a. Upper molars with a trace of the ridges and styles; lower molars with the usual five cusps. Fig. 83. *Choeronycteris mexicana.*

Figure 83

Total length about 77 mm.; forearm about 44 mm.; greatest length of skull, 30.3; zygomatic breadth 10.5. Skull with rostrum very long; tail about half as long as femur, extending less than half way to the very wide interfemoral membrane; calcar distinct but weak; nose-leaf well developed.

Figure 84

3b. Upper molars without ridges and styles; lower molars with cusps strictly lateral (Fig. 84). *Artibeus*. With further characters as follows. Fig. 85. *Artibeus jamaicensis*.

Figure 85

Total length about 70 mm.; forearm about 52 mm.; color brown to gray; nose-leaf well developed; tail does not extend beyond interfemoral membrane. Found rarely.

Family MOLOSSIDAE

This family is easily recognized by the fact that the tail extends fully half its length, or even more, beyond the interfemoral membrane. These bats are known as "free-tailed bats." The wings are long and narrow, thick, and leathery; legs short and stout; nostrils usually have protuberances on them; no tragus present. Found only in the southern part of the United States.

Key to the Genera and Species of *Molossidae*

1a. Total length more than 140 mm., no horny protuberances on anterior portion of ear. *Eumops*. And with the following characters. Fig. 86. *Eumops perotis*. Greater Mastiff Bat.

Total length 157 to 184 mm.; tail 52 to 70 mm.; hind foot 13 to 19 mm.; forearm 69 to 74 mm.; color light brown, with the fur whitish at the base; ears large, directed outward and forward; upper lip with a thick fringe of downward directed hairs; tragus very small; tail extends far beyond interfemoral membrane.

Found in the valleys of southern California and Arizona. A related species, *Eumops glaucinus*, occurs rarely in southern Florida. It is very similar to the Greater Mastiff bat, but is smaller.

Figure 86

1b. Total length less than 140 mm.; a row of horny protuberances on the anterior margin of the ear. *Tadarida*: With further characters as follows. Fig. 87. *Tadarida brasiliensis*. Brazilian Free-tailed Bat.

Total length 90 to 103 mm.; tail 31 to 40 mm.; hind foot 8 to 12 mm.; forearm 40 to 44.; color brown to light brown; ears broad, appearing to be united at base with a series of wartlike projections on anterior border; tragus small and flat; tail long, and about half extends beyond the interfemoral membrane.

This bat is one of the most common in the southwestern states. Great numbers of them inhabit caves like Carlsbad Caverns.

Figure 87

Two related species occur in the United States: *Tadarida femorosacca* is found in southern Arizona and southern California; *Tadarida molossa* occurs throughout the southwest, with one record in British Columbia.

Figure 88

Family VESPERTILLIONIDAE

This is the common family of bats in the United States. The bats in this family are rather small, have big ears with a prominent tragus; the tail is long, but enclosed completely or nearly so in the interfemoral membrane; there is no nose-leaf; molars have a distinct "W" shaped pattern of the cusps.

Key to Genera and Species of *Vespertillionidae*

1a. Lower incisors 4; horse-shoe shaped ridge on snout. *Antrozous*. And with characters as follows. Fig. 89. *Antrozous pallidus*. Pallid Bat.

Total length 99 to 125 mm.; tail 39 to 49,; hind foot 11 to 16 mm.; forearm 46 to 56 mm.; color yellowish brown to light gray; muzzle with a horseshoe-shaped ridge above each nostril, with a large flat swelling behind it; ears separate and wide apart at the base, much longer than the muzzle when laid forward; tragus long, straight and slender, half as long as ear.

Figure 89

This bat lives commonly in the valleys, often invading buildings in hordes, but occuring in caves as well. There are several subspecies, those along the Pacific coast being darker than inland races.

1b. Lower incisors 6; no horse-shoe shaped ridge on snout..........2

2a. Dorsal surface of interfemoral membrane at least partly furred..3

2b. Dorsal surface of interfemoral membrane naked..............5

3a. Total length 90 to 107 mm.; tail 34 to 44 mm.; hind foot about 10 mm.; general coloration blackish with upper parts silvery. *Lasionycteris*. With characters as follows. Fig. 90. *Lasionycteris noctivagans*. Silver-haired bat.

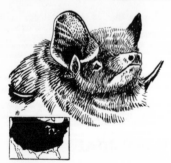

Total length 92 to 107 mm.; tail 34 to 44 mm.; hind foot 10 mm.; forearm 37 to 42 mm.; color blackish brown with many of the hairs tipped with white, imparting a silvery appearance; ears short and broad, about 15 mm. long; when laid forward they do not reach the end of the muzzle.

Figure 90

This bat occurs in the higher areas of America—not in the low valleys. It hides during the day in cracks of trees, under bark, or even in caves. They are most commonly found around water at night.

3b. Total length 125 to 145 mm.; tail 50 to 65 mm.; hind foot about 11
mm.; general coloration brown..............................4

4a. Interfemoral membrane densely furred; teeth 32; ears short and
round. *Lasiurus*. With characters as follows. Fig. 91. *Lasiurus
cinereus*. Hoary Bat.

Total length 128 to 146 mm.; tail 51 to 62 mm.; hind foot 10 to 12
mm.; forearm 49 to 56 mm.; color yellowish brown with silvery white
tips on the hairs giving the appearance of being hoary or almost
silvery; whole dorsal surface of interfemoral membrane furred;
ears short, about 18 mm. long, almost as wide as high; tragus almost triangular, and about half as long as the ear.

The hoary bat lives commonly in
northern regions, about lakes and
over meadows. It is one of the largest of our bats, and can often be distinguished on the wing by the large
size and narrow wings. Its flight is
erratic, making it a difficult target.
Most specimens are taken from trees
by fruit pickers, for the bats hang
there during the day.

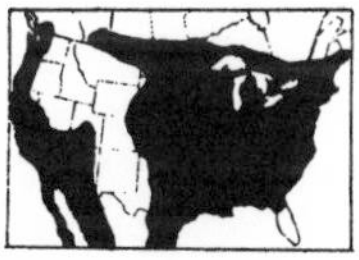

Figure 91

The Red Bat, *Lasiurus borealis*, (fig. 92) resembles
the hoary bat, except that the color is bright reddish or even rusty red with the tips of hairs whitish, giving a frosted appearance; buffy white
patch on each shoulder. Total length 95 to 112
mm.; tail 45 to 62 mm.; hind foot 8½ to 10 mm.;
forearm about 40 mm. Habits similar to those of
the Hoary Bat..

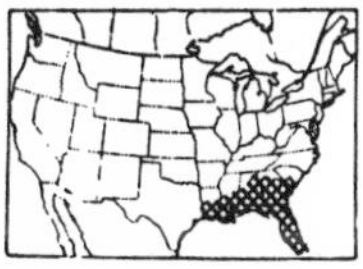

Figure 92

A third species the Seminole Bat, *Lasiurus seminolus*,
(fig. 93) is similar to the Red Bat above, but much
darker in color, about mahogany, with less frost-
ing on the hair tips.

Figure 93

4b. Interfemoral membrane furred to the base only; teeth 30; ears longer than broad. *Dasypterus*. With characters as follows. Fig. 94. *Dasypterus floridanus.* Florida Yellow Bat.

Figure 94

Total length 115 to 129 mm.; tail 50 mm.; hind foot 10 mm.; color light yellowish brown to reddish brown; ears medium, but broad and rounded; tragus triangular; interfemoral membrane haired for about the basal third on dorsal surface.

Figure 95

This bat probably does not rest in caves, but hangs in trees, so it is not often found. A related species, *Dasypterus intermedius*, (fig. 95) occurs also in the gulf states; it is larger, total length 145 mm.; tail 65 mm.

5a. Length of ear half, or more than half, the length of the forearm...6

5b. Length of ear less than half the length of forearm. Fig. 96. *Nycticeius humeralis.* Twilight Bat.

Figure 96

Total length 93 mm.; tail 37 mm.; hind foot 7 mm.; forearm 35 mm.; color dull brown; teeth 30 in number; ears small, thick and leathery; skull short, broad and low.

This bat hides during the day in a hollow tree, then comes out in the evening to catch insects on the wing.

6a. Length of ear from notch about 28 to 35 mm. 7

6b. Length of ear less than 25 mm. 8

7a. Tragus broadly rounded at tip; back blackish with three white patches. *Euderma.* With the following characters. Fig. 97. *Euderma maculatum.* Spotted Bat.

Figure 97

Total length 105 to 110 mm.; tail 47 to 50 mm.; hind foot 9½ to 10 mm.; forearm 45 to 50 mm.; ear about 32 mm.; color blackish with three whitish patches on the back, one on each shoulder and one on the rump; ears very large and joined across the forehead by a narrow membrane; tragus long, about 19 mm.

7b. Tragus slender, pointed at tip; color not blackish, and no white patches present. Fig. 98. *Plecotus townsendii.* Lump-nosed Bat.

Figure 98

Total length 92 to 108 mm.; tail 43 to 55 mm.; hind foot 8 to 12 mm.; forearm 37 to 43 mm.; ear 29 to 36 mm.; color wood brown to light brown; ear very long; tragus slender, almost half as long as the ear; nostrils very irregular, with lumpy growths around them.

These bats are veteran cave dwellers, hanging in small clusters from a high ceiling. Since they remain alert at all times, one must be skillful in order to capture them in a net They are also found in attics of old buildings. A related species, *Plecotus rafinesquii,* (fig. 99) is slightly darker in color; it occurs in the southern states.

Figure 99

8a. Tragus short, blunt, and curved forward; size very small, only about 65 to 75 mm. Fig. 100. *Pipistrellus hesperus.* Little Canyon Bat.

Figure 100

Total length 62 to 80 mm.; tail 24 to 30 mm.; hind foot 5 mm.; forearm 26 to 30 mm.; color buffy gray above, and whitish below; membranes, ears, and feet blackish; tragus blunt with tip bent forward.

This bat in the west lives only in rocky canyons where it hides in cliffs during the day. It goes first to a stream for a drink, then flies out for insects for the next few hours. Specimens killed only a few minutes after emerging in the evening have had the stomachs completely filled with insects.

There is another representative of this genus, the pipistrelle or Pygmy Bat. *Pipistrellus subflavus* (fig. 101).

Figure 101

Total length 81 to 89 mm.; tail 36 to 45 mm.; hind foot 8 to 10 mm.; forearm 33 mm.; similar to the species above, except that the tragus is tapering and straight instead of blunt and bent forward; size larger than the species above. Found over nearly all the eastern states.

8b. Tragus longer, slender, and not curved forward...............9
9a. Total length more than 105 mm.; teeth 32. Fig. 102. *Eptesicus fuscus.* Big Brown Bat.

Figure 102

Total length 107 to 112 mm.; tail 38 to 47 mm.; hind foot 8 to 10 mm.; forearm 43 to 45 mm.; color reddish brown to pale wood brown; ear about 18 mm. long, about 8 mm. wide; tragus about 12 mm., tapering and tipped slightly forward; membranes nearly naked.

This bat rests during the day in buildings, in trees or in caves, coming forth later in the evening than do most other bats. It often is found to remain active later in the fall than do most other bats, then it hibernates in buildings and caves. Several subspecies occur in the United States.

9b. Total length less than 105 mm.; teeth 38. Fig. 103. Myotis. Common Bat.

Total length seldom over 95 mm., never over 103 mm.; tail rather long; ear oval, higher than wide; tragus slender and rather straight; wings broad; fur soft and silky, never bright in color, ranging from dark brown through reddish brown to buff in species of arid regions.

Figure 103

These bats are insect-eating, and may be found any summer evening flying over mountains and valleys in search of insects. They spend the day in caves, attics, or under loose bark of trees. Most species hibernate in winter, but a few migrate. About 30 species and subspecies occur in our country. Since these bats are so difficult to identify as to species, I have chosen to omit them from this general key. For anyone who may wish it, I will include a non-illustrated technical key to these species.

Key to the Species of Myotis

1a. Underside of wing furred to level of elbow; skull with rostrum shortened and occiput unusually elevated; found in western North America. Myotis volans.

1b. Underside of wing not furred to level of elbow; skull with normal rostrum and occiput..2

2a. Foot small, the ratio of its length to that of tibia ranging from about 40 to 46 3

2b. Foot normal or large, the ratio of its length to that of the tibia ranging from about 48 to 60..................................4

3a. Hairs of back with long shiny tips; third metacarpal not so long as forearm; skull larger, with flattened braincase and gradually rising profile; entire United States. Myotis subulatus.

3b. Hairs of back dull-tipped; third metacarpal usually as long as forearm; skull smaller, with rounded braincase and abruptly rising profile; western United States. Myotis californicus.

4a. Wing membrane attached to tarsus; fur of back without obvious darkened basal area; ratio of foot to tibia usually about 60; from Indiana and Illinois to Alabama and Georgia. Myotis grisescens.

4b. Wing membrane attached to side of foot; fur of back with obviously darkened basal area; ratio of foot to tibia usually less than 57 5

5a. Fur of back with an obvious tricolor pattern, calcar usually with a small but evident keel; eastern United States. *Myotis sodalis.*

5b. Fur of back without an obvious tricolor pattern; calcar normally with no trace of keel...6

6a. Ear when laid forward extending noticeably beyond tip of muzzle 7

6b. Ear when laid forward not extending noticeably beyond tip of muzzle 10

7a. Free border of uropatagium with inconspicuous, scattered, stiff hairs; central and northern North America. *Myotis keenii.*

7b. Free border of uropatagium usually with a noticeable fringe of stiff hairs 8

8a. Size larger; forearm usually 41 to 46 mm.; ear not very large; fringe conspicuous; western states and Mexico. *Myotis thysanodes.*

8b. Size smaller; forearm usually 33 to 40 mm.; ear very large; fringe not conspicuous; western states and northern Mexico............9

9a. Skull with noticeably flattened brain case; forearm 33 to 36 mm.; Lower California. *Myotis milleri.*

9b. Skull with normal braincase; forearm 37 to 40 mm. *Myotis evotis.*

10a. Cheek teeth robust, the breadth of the maxillary molars, as compared with that of the intervening palate, greater than usual in American members of the genus............................11

10b. Cheek teeth normal, the breadth of the maxillary molars as compared with the intervening palate not greater than usual in American members of the genus................................12

11a. Brain case flattened; fur glossy; southwestern states. *Myotis occultus.*

11b. Brain case highly arched; fur dull in color; western states, Kansas through Texas to New Mexico. *Myotis velifer.*

12a. Fur above dense, wooly; a low but evident sagittal crest always present in adults; Florida, Indiana. *Myotis austroriparius.*

12b. Fur above normal, silky; sagittal crest rarely present.........13

13a. Forearm ranging from 32 to 37; greatest length of skull 13.2 to 14.2 mm.; hairs of back without conspicuous burnished tips; western states. *Myotis yumanensis.*

13b. Forearm ranging from 36 to 40 mm.; greatest length of skull ranging from 14.3 to 15.3 mm.; hairs of back with conspicuous burnished tips; all of North America. *Myotis lucifugus.*

Order LAGOMORPHA
Rabbits, Hares, Pikas

This order is familiar to everyone, for rabbits and hares occur in every part of the country. One family, OCHOTONIDAE, the pika or cony family, is not so well known, although anyone who visits the high mountains of the west is sure to see these little rock rabbits perched on top of a boulder on a rocky mountainside.

Key to the Families of LAGOMORPHA

1a. Tail present; hind legs developed for jumping; ears much elongated; size larger than small squirrel-size. LEPORIDAE, page 70.

1b. Tail absent (except for a remnant which is not visible until after the animal is skinned); legs not developed for jumping; ears short and rounded; size about that of a small squirrel. OCHOTONIDAE, page 69.

Family OCHOTONIDAE
Pikas

Pikas live in the high mountains in rock slides where the boulders are usually quite large, and where there is little or no growth of plants among the boulders. Plants are always found in abundance near the rock slides, for the pikas feed entirely upon green plants. These they cut during the summer months, carry them to the rocks, and make hay stacks among the boulders to dry. When dry the hay is carried under the rocks where the pikas feed on the dried plants throughout the winter. They never hibernate, but are active during the winter months. In the high mountains, where they ordinarily occur, they spend 8 or 9 months under the snow. Sometimes they are found at low elevations, as along the Columbia River in Oregon and Washington, but this is not typical.

Collecting pikas is no easy job unless one wishes to shoot them. Since they feed on green plants it is almost impossible to trap them, although one mammalogist had success with dried prunes.

Pikas are active in the day time, and one may see the little fellows perched on top of a boulder sunning themselves. Their ears are keen, and it is difficult to come very near to them. Since all the subspecies have now been placed in one species, no key need be included.

Fig. 104. *Ochotona princeps.* **Pika, Cony, or Rock Rabbit.**

Figure 104

Total length, 180 to 200 mm.; tail lacking externally, but usually 5 to 7 mm. when skinned out; hind foot 29 to 33 mm.; color grayish or buffy brown, sometimes darker brown.

Pikas are so isolated among the mountains of the western states that a large number of races or subspecies occur. These vary in coloration, size, and cranial characters. Habits are usually about the same for all the subspecies.

Family LEPORIDAE
Rabbits and Hares

Rabbits and hares occur in almost every habitat in our country. In the deserts of the west the long-legged, long-eared jack rabbits are a familiar spectacle. In the forests of the mountains the snowshoe rabbits or varying hares hop over the snow, leaving their huge "snowshoe" tracks. The cottontails love the brushy areas in the proximity of man. The swamp rabbits prefer the marshy areas of the south. The white-tailed jackrabbit roams the plains from the central states to Canada.

Key to Genera and Species of RABBITS

1a. Size large, 400 mm. or more, often much more; hind feet more than 125 mm.; ear 100 mm. or more. **Lepus**........................2

1b. Size smaller, 400 mm. or less, often around 300 to 350 mm.; hind foot less than 125 mm.; ear 40 to 70 mm. **Sylvilagus**............8

2a. Color white throughout the year. *Lepus arcticus.* **Arctic Hare.**
Total length about 600 mm.; tail 60 to 70 mm.; hind foot 165 mm.;
ear 80 mm.; color all white in winter, except for the black tips on
the ears; in summer the color is grizzled white.

The Arctic Hare indeed inhabits only the arctic regions, so you
must go there to see him, unless you can find a specimen in the zoo.

2b. Color either not white at all, or else white only in the winter....3

**3a. Total length 500 mm. or more; tail 75 mm. or more; ear nearly as
long as hind foot; found in arid and grassland areas. Jackrabbits..4**

**3b. Total length less than 500 mm., usually from 400 to 450 mm.; tail
less than 75 mm., usually about 50 mm.; ear much.shorter than**

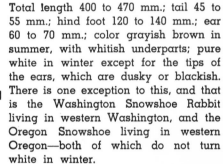

hind foot; found most commonly in timbered areas. Fig. 105 (winter phase).
Fig. 106 (summer phase). *Lepus americanus.* **Snowshoe Rabbit or Varying
Hare.**

Total length 400 to 470 mm.; tail 45 to
55 mm.; hind foot 120 to 140 mm.; ear
60 to 70 mm.; color grayish brown in
summer, with whitish underparts; pure
white in winter except for the tips of
the ears, which are dusky or blackish.
There is one exception to this, and that
is the Washington Snowshoe Rabbit
living in western Washington, and the
Oregon Snowshoe living in western
Oregon—both of which do not turn
white in winter.

Figure 105

Figure 106

The snowshoe rabbit lives in the forests of America where winter snows
become very deep. It feeds on a variety of plant food. The animal's large
hind feet enable it to hop over soft
snow with ease, and its white winter
fur protects it from the keen eyes of
predatory animals and birds. A number of subspecies occur throughout the
northern states.

4a. Tail whitish above with a narrow grayish stripe down the center, pure white below; no yellow area on flanks; found mainly in grassland regions; fur turns white in winter. Fig. 107. *Lepus townsendii.* **White-tailed Jackrabbit.**

Figure 107

Total length 550 to 600 mm.; tail 75 to 115 mm.; hind foot 135 to 155 mm.; ear 100 to 110 mm.; color grayish in summer, somewhat grizzled by whitish hairs among the gray. In winter the fur is nearly pure white, especially in the northern part of its range; farther south the color may be often grizzled with both white and gray.

The white-tailed jackrabbit is mainly a prairie rabbit, living all its life in the grassland areas of the prairie states or in grassy uplands in other regions.

4b. Tail black above and gray below; yellow or whitish area on flanks; fur does not turn white in winter; found mainly in arid regions.....5

5a. Color on flanks whitish, distinctly differing from that of the back; or else belly pure white.....................................6

5b. Color of flanks like that of the back. Fig. 108. *Lepus californicus.* **Black-tailed Jackrabbit.**

Figure 108

Total length 500 to 550 mm.; tail 70 to 105 mm.; hind foot 120 to 130 mm.; ear 120 to 160 mm.; color gray or grizzled gray, often with a buffy cast; top of tail and ear tips black; underside of tail buffy gray.

This is the most common jackrabbit on the plains and uplands of the western states. It is especially common in arid regions, and becomes a pest near farms, for it does much damage to the crops. A number of subspecies occur, but all are similar in size and color. The trails of this rabbit are common sights among the sagebrush, and extend over the hills in a regular network.

6a. Length 700 mm. or more; no white on flanks. *Lepus europaeus.* European Hare.

Total length 700 to 760 mm.; tail 75 to 80 mm.; hind foot 135 to 150 mm.; color yellowish brown with grayish cheeks and rump; underparts pure white; tail black above and white below; skull very large, 90 to 105 mm. in length.

This European rabbit has been introduced in Ontario, Canada, and has migrated south into Michigan, Minnesota and Wisconsin. In some areas it has escaped from captivity, as in some of the San Juan Islands of Washington.

6b. Length 600 mm. or less; color on flanks whitish, distinctly differing from that of the back.....................................7

7a. Size very large, length about 600 mm.; ears enormous, about 150 mm.; sides of body and rump iron gray. Fig. 109. *Lepus alleni.* Antelope Jack Rabbit.

Tail about 65 mm.; hind foot 130 mm.; legs very long and slender; tail very short and small; color gray on sides, rump, shoulders and flanks; head buff; top of back buff, often with a pinkish tinge.

This is the most strikingly colored jack rabbit in America, and is also the fastest. It is the largest as well, except for the White-tailed Jack Rabbit. It inhabits the deserts.

Figure 109

7b. Size smaller, about 500 to 550 mm.; ears medium, 115 to 120 mm.; flanks white; rump iron gray. Fig. 110. *Lepus gaillardi.* Gaillard Jack Rabbit.

Tail 70 to 80 mm.; hind foot 120 to 135; upper parts rich buff; white area extends from shoulder to rump and from abdomen to flanks; outside of rump and thigh gray.

This rabbit is able (as is the Antelope Jack Rabbit above) to move the white area of the sides upward or downward by a series of muscles under the skin. When the animal runs it moves this area alternately up and down, or may keep the white area facing the observer, and concealing the darker colors. No one knows why the rabbit behaves this way.

Figure 110

8a. Tail rather large, loosely haired and cottony white on the underside (underside usually conspicuously visible, for the tail is held upright) **9**

8b. Tail rather short, densely haired and white, yellowish or gray beneath **13**

9a. Occurs east of the Mississippi River.......................... **10**

9b. Occurs west of the Mississippi River......................... **11**

10a. Black patch between ears; color pinkish buff overlaid with blackish; tail only 38 to 40 mm.; total length less than 400 mm.; ear only about 50 mm.; supraorbitals very small, posterior process short, tapering posteriorly to a slender point, free from or barely touching skull and narrowing anteriorly. Fig. 111. *Sylvilagus transitionalis.* New England Cottontail. Hind foot 95 mm.; color pinkish buff; narrow black patch between ears; sides often grayish; back overlaid with a blackish wash.

Figure 111

The New England Cottontail is a brush or woods-loving species and stays hidden most of the time. It feeds on grasses and herbaceous plants, often for some time after sunrise.

10b. Reddish patch behind ears; color buffy brown washed with gray and sprinkled with black; tail usually more than 50 mm., often 55 to 65 mm.; total length usually more than 400 mm; ear usually more than 50 mm., often about 60 mm., supraorbitals broadly developed; posterior process usually broadly strap-shaped and coalescing with skull posteriorly and sometimes along entire length. Fig. 112. *Sylvilagus floridanus.* Eastern Cottontail.

Total length 375 to 465 mm.; tail 39 to 65 mm.; hind foot 87 to 104 mm.; ear 49 to 67 mm.; color of upper parts varies from dark grayish buff to buffy brown, often with blackish hairs mixed in with the buff; underparts pale buff.

Figure 112

These are the most common rabbits throughout the eastern states, and much of the great plains. They prefer brushy areas, swampy places, or patches of weeds about farms. In fact, they seem to live almost anywhere that they can get adequate shelter from their enemies, but are not often found in heavy timber. A number of subspecies occur.

11a. Ears usually more than 65 mm. long. Fig. 113. *Sylvilagus auduboni*. Western Cottontail.

Total length 359 to 418 mm.; tail 45 to 72 mm.; hind foot 76 to 97 mm.; ear 59 to 69 mm.; color dark buffy brown to dull grayish buff, depending upon the subspecies and the time of year.

Figure 113

These cottontails inhabit the western states in much the same manner as do the eastern cottontails in the east. The animal is somewhat smaller, differs in several cranial characters, and has slightly longer ears than its eastern relative.

11b. Ear 50 to 60 mm..12

12a. Rostrum proportionately heavy, broad and strongly angled on upper half of base, usually broad and flattened, or decurved, near tip; supraorbitals broad and heavy. *Sylvilagus floridanus*. Eastern Cottontail. See description under 10b.

12b. Rostrum long and slender, not strongly angled on upper half of base; outlines straight, narrow and rounded at tip; supraorbitals always light and slender, tapering to a narrow point. Fig. 114. *Sylvilagus nuttallii*. Washington Cottontail.

Total length 325 to 425 mm.; tail 36 to 50 mm.; hind foot 85 to 100 mm.; ear 52 to 68 mm.; color dark buffy brown, the sides somewhat lighter, or even grayish; underparts whitish with the hind legs rusty; skull with a narrow rostrum and with bullae smaller than in other cottontails.

Figure 114

This is the common cottontail of the Great Basin and the northwestern states, except where eastern cottontails have been introduced. The habits of this rabbit are similar to those of the Eastern Cottontail.

13a. Size very small, about 280 to 300 mm.; tail very short, about 18 to 20 mm.; hind foot only 65 to 75 mm.; ear only 40 to 42 mm.; color of tail gray or buff. Fig. 115. *Sylvilagus idahoensis*. Idaho Pygmy Rabbit.

Figure 115

Color dull brownish gray; sides grayish; underparts whitish; skull short and wide, with large bullae.

The Pygmy Rabbit is the smallest of the cottontails, although its tail is not white. It occurs in arid regions, where it is usually not very common.

13b. Size larger in total length, tail, foot, and ear; tail gray, buff or white **14**

14a. Underside of tail white.....................................**15**

14b. Underside of tail gray or buff. Fig. 116. *Sylvilagus palustris.* Marsh Rabbit.

Total length 425 to 440 mm.; tail 30 to 40 mm.; hind foot 85 to 95 mm.; ear 40 to 55 mm.; color buffy grayish brown, often rusty or reddish brown, but also sometimes grayish; underparts reddish brown except for the white area in center of abdomen.

Figure 116

This little rabbit lives right among the marsh plants in the swamps along the South Atlantic coast, and feeds upon green plants. It swims with ease, and is able to walk like a dog or cat, moving one foot at a time, although it can hop like other rabbits. It is smaller than the Swamp Rabbit below.

15a. Size less than 400 mm.; tail small and round with short dense fur; occurs in the coastal area of California and Oregon. Fig. 117. *Sylvilagus bachmani.* Brush Rabbit.

Total length 310 to 360 mm.; tail 25 to 40 mm.; hind foot 70 to 80 mm.; ear 49 to 63 mm.; color dark reddish brown fading to grayish brown in worn pelage.

Figure 117

These little rabbits live in the dense undergrowth of the coastal areas where timber has been cut off from the hills and mountains along the coast. They are seldom seen because of their seclusive habits.

15b. Size more than 500 mm.; tail slender and thinly haired; occurs from Texas to Georgia, and north along the Mississippi to southern Illinois. Fig. 118. *Sylvilagus aquaticus.* Swamp Rabbit.

Total length 525 to 540 mm.; tail 65 to 75 mm.; hind foot 100 to 110 mm.; ear 62 to 70 mm.; color yellowish brown, darker than the common cottontail, often being reddish brown; underparts white; head large, size noticeably larger than the Marsh Rabbit.

Figure 118

Habits very much like those of the Marsh Rabbit, for it lives in the wet bottomlands, marshy areas along rivers. Here it makes trails through the swampy ground, walking rather than hopping. They are excellent swimmers.

Order RODENTIA

Squirrels, Gophers, Mice, Porcupines, Beavers

Rodents can be recognized at once by their long chisel-like incisors, the lack of canine teeth, and the presence of rather large molars with wide grinding surfaces. The absence of canine teeth causes a wide space (the *diastema*) between the incisors and the premolars. In size they vary all the way from the smallest mouse to the beaver, while even larger rodents occur in South America. Rodents are the most abundant of the mammals in America, and can be found in almost every habitat from the driftwood strewn beaches to the rock slides of the high mountains. Members of this order are so diverse in their habits that little can be said about them in general except that they feed largely on plant food.

Key to the Families of RODENTIA

1a. Upper parts of body, especially the lower back and tail, partially covered with stiff spine-like quills; infraorbital foramen enormous. *ERETHIZONTIDAE*, porcupines, page 156.

1b. Upper parts of body not covered with quills; infraorbital foramen not unusually large 2

2a. External cheek pouches present 3

2b. External cheek pouches absent 4

3a. Tail either as long as, or longer than head and body; forelegs somewhat reduced in size; skull widened posteriorly with large paper-thin bullae; claws of forefeet not greatly enlarged. HETERO-MYIDAE, pocket mice and kangaroo rats, page 114.

3b. Tail much shorter than head and body; forelegs not reduced; skull not widened posteriorly, and without large paper-thin bullae; claws of forefeet greatly enlarged. GEOMYIDAE, pocket gophers, page 108.

4a. Hind legs adapted for jumping; forelegs reduced; tail much longer than head and body. ZAPODIDAE, jumping mice, page 154.

4b. Hind legs not adapted for jumping; forelegs not reduced; tail shorter 5

5a. Body large and heavy, 30 inches or more in length..............6

5b. Body smaller 7

6a. Tail wide and flattened dorso-ventrally. CASTORIDAE, beavers, page 127.

6b. Tail not wide and flattened, but round. CAPROMYIDAE, coypus, page 157.

7a. Tail usually scaly and sparsely haired; teeth 16...............9

7b. Tail not scaly, hairy and often bushy; or tail vestigial, scarcely visible at all; teeth 20 or 22.................................8

8a. Tail vestigial; skull strongly flattened, with flask-shaped auditory bullae; no postorbital spines on frontal bones. APLODONTIDAE, mountain beavers, page 78.

8b. Tail usually long and often bushy; skull not so strongly flattened and without flask-shaped auditory bullae; postorbital spines present on frontal bones. SCIURIDAE, squirrels, page 79.

9a. Maxillary molars with three longitudinal series of tubercles. MURIDAE, European mice and rats, page 153.

9b. Maxillary molars with two or less longitudinal series of tubercles, or with complicated folds of enamel without any tubercles. CRICETIDAE, American mice and rats, page 128.

Family APLODONTIDAE
Mountain Beavers

Mountain beavers live only in the coastal area from British Columbia south to northern California. They are large, heavily built rodents with extremely short tails—almost no tail at all. The head is very large; front feet and front legs are adapted for digging. These animals dig extensive tunnels just under the surface in wooded areas, much like pocket gophers or moles, but on a much larger scale. They feed on roots of many plants, even eating the rhizomes of ferns. Very little is known of their life history, and nearly all attempts to keep them in captivity have failed.

Fig. 119. *Aplodontia rufa.* **Mountain Beaver.**

Figure 119

Total length 325 to 440 mm.; tail 28 to 40 mm.; hind foot 48 to 65 mm.; color grayish brown to brownish gray, grizzled with numerous blackish hairs; underparts gray, paler than upper parts.

Only one species of *Aplodontia* occurs in America, but there are several subspecies. All the subspecies are so nearly alike that it is very difficult to distinguish them. Mountain beavers live generally in mountain forests, near water, but they also occur in the timbered valleys of western Washington, western Oregon, and northern California.

Family SCIURIDAE
Squirrels

The squirrel family includes not only the tree squirrels which are so common in the wooded areas and in city parks, but also the chipmunks, flying squirrels, woodchucks and ground squirrels. The ground squirrels are called "gophers" by many people, but this name should only be applied to the pocket gopher in the family GEOMYIDAE. Most squirrels have long tails, often bushy; some of the ground squirrels have rather short tails, and seldom are they bushy.

Key to the Genera of SCIURIDAE

1a. Tail short and bushy; body stout and heavy....................2

1b. Tail, if short, not bushy; body rather slender and light in weight..3

2a. Length 400 mm. or less; color usually light buff with no contrasting colors of any kind. *Cynomys,* prairie dogs, page 107.

2b. Length over 600 mm., color usually brownish or grayish brown, but always with contrasting colors, especially about the nose and mouth. *Marmota,* marmots or woodchucks, page 84.

3a. Tail flattened horizontally due to growth of fur on lateral portions; patagium present. *Glaucomys*, flying squirrels, page 105.

3b. Tail not flattened; no patagium present......................4

4a. Tail long and bushy. The tree squirrels......................5

4b. Tail, if long, not often bushy. The ground squirrels and chipmunks 6

5a. Color of underparts contrasting sharply with that along the sides and upper parts; black line along sides separating the upper parts from the underparts; anterior upper premolar vestigial or absent; baculum absent. *Tamiasciurus*, the red squirrels, pine squirrels, or chickarees, page 83.

5b. Color of underparts not contrasting sharply with that along the sides and upper parts; no black line along sides; anterior upper premolar well developed; baculum present. *Sciurus*, the tree squirrels, page 80.

6a. Longitudinal light and dark stripes present on back and on sides of head; feet adapted for climbing; antorbital canal absent, the antorbital foramen piercing the zygomatic plate of the maxillary...7

6b. Longitudinal stripes usually absent, but if present, then not found on the head; feet not adapted for climbing; antorbital canal present. *Citellus*, ground squirrels, page 87.

7a. Back with 4 light stripes; stripes continue to base of tail; 5 cheek teeth in upper jaw. *Eutamias*, western chipmunk, page 97.

7b. Back with only 2 light stripes; stripes do not continue to base of tail; only 4 cheek teeth in upper jaw. *Tamias*, eastern chipmunk, page 96.

Genus SCIURUS
Tree Squirrels

These squirrels are usually large, have long bushy tails, pointed and frequently tufted ears, with the feet adapted for climbing. The fore feet have four toes, and the hind feet five toes. They usually nest in trees, making the nest in a hollow tree or else constructing a large ball of bark shreds and other plant fibers and placing it on a horizontal limb often high in a tree. They do not hibernate, but sleep during stormy weather, coming out as soon as the weather is good. They store food in caches near the nest.

Key to the Species of Tree Squirrels

1a. Upper premolars two, molars three, making five grinding teeth on each side of upper jaw......................................2

1b. Upper premolar one, molars three, making four grinding teeth on
each side of upper jaw...5

2a. Ears tufted with long black hairs; color reddish brown to grayish
brown, but with a distinct reddish area in middle of back........3

2b. Ears not tufted; color grayish to silvery gray; no reddish color on
back 4

3a. Belly white; ears black; reddish area small, and in center of back.
Fig. 120. *Sciurus aberti*. Abert Squirrel.

Figure 120

Total length about 525 mm.; tail about
225 mm.; hind foot about 75 mm.; color
reddish brown to grayish brown with
a small patch of bright reddish color
in middle of back; belly white, and separated from the brownish or grayish
color of upper parts by a black line
along the sides; ears very long and
tufted. Two subspecies of this squirrel
found in Colorado and northern New
Mexico lack the reddish patch in center
of back, but are otherwise similar.

The Abert Squirrel is one of the most beautiful animals in America.
It can be found in the higher mountains of its range, but is rather
wary due to much hunting.

3b. Belly dark or even black; ears reddish; reddish area of back very
large, almost covering the entire back. Fig. 121. *Sciurus kaibab-
ensis*. Kaibab Squirrel.

Figure 121

Measurements about the same as
those for the Abert Squirrel. Colors
differ in that the belly is black, ears
reddish, and the reddish area of the
back is much larger.

This is the most beautiful of all the
squirrels. In the Grand Canyon National Park the Kaibab Squirrel has
become very friendly, where it is
rigidly protected by law. It occurs
only on the north rim of the Canyon.

4a. Upper parts silvery gray due to white-tipped hairs; tail very long and bushy; total length about 550 to 585 mm.; tail about 250 to 285 mm.; found only in the Pacific Coast States. Fig. 122. *Sciurus griseus.* Western Gray Squirrel, or Silver Gray Squirrel.

Hind foot 80 to 85 mm.; ear about 30 mm.; weight about 2 pounds; color silvery gray on the upper parts; underparts white; tail all gray with white tips on the long hairs.

This beautiful squirrel is still common in the mountains of northern California, and in the Sierra Nevadas, but is rather rare in the Cascades of Oregon and Washington. Since it is a game animal in the latter two states it may eventually be exterminated there unless sentiment for its protection can be aroused.

Figure 122

4b. Upper parts mixed gray and yellowish brown, with the head and back darker and more brownish; under parts white; tail appears blackish overlaid with white; total length 450 to 485 mm.; tail about 200 mm.; the most common tree squirrel in the eastern states. Fig. 123. *Sciurus carolinensis.* Eastern Gray Squirrel.

This beautiful squirrel is the most familiar tree squirrel in America. In the eastern states it is native, and in the west it has been introduced into city parks until now it has become well adjusted to western habitats, even spreading out into the wooded areas outside the cities.

Figure 123

5a. Occurs in Arizona and New Mexico only.......................6

5b. Occurs only in the central and eastern states, south to Texas, although it has been introduced in cities in many other states. Fig. 124. *Sciurus niger.* Fox Squirrel.

Total length about 515 to 550 mm.; tail about 250 mm.; hind foot about 70 mm.; color tawny brown grizzled with gray on the upper parts; underparts pale yellowish brown or even reddish brown; tail very long and bushy, a mixture of black and reddish brown; ears rather short for such a large squirrel.

Figure 124

6a. Color reddish brown above and yellowish to reddish below; tail a mixture of black and reddish brown; found only in southern Arizona, then south into Mexico. Fig. 125. *Sciurus apache.* Apache Fox Squirrel.

This squirrel is similar to *Sciurus niger.* It inhabits brushy areas and timbered regions at moderate heights in the mountains.

Figure 125

6b. Color grizzled gray, not reddish above; underparts white; tail black above and brown below, with white tips on the hairs; found in the mountains of Arizona and New Mexico. Fig. 126. *Sciurus arizonensis.* Arizona Fox Squirrel.

Figure 126

Genus TAMIASCIURUS
Red Squirrels

These squirrels may be distinguished from the gray squirrels and other tree squirrels by their coloration and smaller size. If one spends some time watching both types of squirrels, it will not be long until he can tell them apart readily.

Key to the Species of Red Squirrels

1a. Belly white or nearly white.....................................2

1b. **Belly yellow, orange, or rusty. Fig. 127.** *Tamiasciurus douglasii.* **Douglas Squirrel or Chickaree.**

Figure 127

Total length about 300 mm.; tail 115 to 140 mm.; hind foot about 50 mm.; ear about 25 mm.; color dark reddish brown; black stripe along sides; underparts varying from dark orange to pale yellowish depending upon the subspecies; tail reddish brown tipped and edged with orange or with white.

This squirrel inhabits the forests of the coastal area and the timbered regions in the Cascades and Sierras of the three Pacific states.

2a. **Color of back reddish brown to grayish brown. Fig. 128.** *Tamiasciurus hudsonicus.* **Red Squirrel.**

Figure 128

Total length about 350 mm.; tail 125 to 150 mm.; hind foot about 50 mm.; ear about 25 mm.; color rusty gray to reddish brown; black stripe along the sides; underparts white; tail mainly black.

The Red Squirrel is one of the most familiar mammals in the northern states, becoming a pest in some places. It makes its nest of shredded bark and plant fibers, placing it as high as 40 feet up in a tree.

Figure 129

2b. **Color of back pale yellowish red; otherwise similar to** *Tamiasciurus hudsonicus* **in all respects. Fig. 129.** *Tamiasciurus fremonti.* **Pine Squirrel or Chickaree.**

Genus MARMOTA
Marmots or Woodchucks

Marmots are heavy set animals with short ears, short stubby tails, small cheek pouches, and with the front feet with four toes. The skull is heavy and wide. These animals live in rock slides and boulder piles in the western states; some inhabit the dry hot valleys, while others prefer the high mountains. In the east they will live in meadows and rolling hilly country where there are brushy tangles interspersed with open fields and meadows. They dig burrows into banks, or into the ground, while western marmots live under boulders, seldom digging burrows at all.

Key to the Species of Marmots

1a. Upper parts mainly black and white, cinnamon buff on the rump. Fig. 130. *Marmota caligata*. Hoary Marmot, or Whistler.

Figure 130

Total length 670 to 786 mm.; tail 185 to 250 mm.; hind foot 91 to 112 mm.; color of foreback black and white, the colors rather well mixed so that a silvery or hoary appearance is the result; hindback near rump also mixed black and white, but with a strong tingle of cinnamon buff; underparts white, or grayish white, or even cinnamon or blackish brown.

The Hoary Marmot is a familiar sight to the mountain climber in the northern Rockies, or in the Cascades in Washington. Here it inhabits the boulder slides along with the pikas and golden-mantled ground squirrels. Marmots feed on herbaceous plants, and must sometimes travel some distance to a meadow. They are always on the alert. When a group of marmots are feeding, one will act as sentinel; when danger comes near he will give a shrill whistle, then all the marmots make for the rock slide. Since his eye sight is very poor, the marmot depends on his keen hearing to detect danger. The Hoary Marmot hibernates for nine months or so out of the year, for his rock slide is buried that long in the deep snow.

1b. Upper parts mainly brownish, yellowish, drab, or buff...........2

2a. Upper parts solid colors, not grizzled...........................3

2b. Upper parts grizzled brownish or yellowish brown...............4

3a. Upper parts brownish drab, buffy or reddish brown. Fig. 131.
Marmota olympus. Olympic Marmot.

Figure 131

Total length 710 to 790 mm.; tail 195 to 220 mm.;
hind foot 108 to 112 mm.; color of upper parts
brownish drab mixed with a few whitish hairs; head
blackish brown with a white patch before the eyes;
underparts about the same color as upper parts.

This marmot is very closely related to the Hoary
Marmot, but lacks the silvery color of the back. It
is completely isolated from all other types of marmots, for it lives
only in the Olympic Mountains of Washington state.

3b. Upper parts blackish brown or black. Fig. 132. Marmota vancou-
verensis. Vancouver Island Marmot.

Figure 132

Total length 660 to 710 mm.; tail about 222 mm.;
hind foot 90 to 102 mm.; color dark brown over
entire body. This marmot is closely related to the
Hoary Marmot, but has been isolated for so long
that the dark brown colors have become predomi-
nant.

4a. Sides of neck with conspicuous buffy patches. Fig 133. Marmota
flaviventris. Yellow-bellied Marmot.

Figure 133

Total length 600 to 700 mm.; tail 170
to 180 mm.; hind foot 85 to 90 mm.;
color of upper parts reddish to yellowish brown grizzled with white
hairs; underparts buff to yellowish
buff.

The Yellow-bellied Marmot inhabits the rocky hillsides, highway fills,
railroad grades, and rocky canyons.
It seems to have little regard for
altitude, being found at both low and
high elevations. In low hot valleys
it must go into aestivation in early
summer, for the drying up of the
green plants terminates its food supply.

4b. Sides of neck without conspiciuous buffy patches. Fig. 134. _Marmota monax_. Common Marmot or Woodchuck.

Total length 418 to 675 mm.; tail 105 to 155 mm.; hind foot 75 to 88 mm.; color of upper parts grayish brown to reddish brown grizzled with many white-tipped hairs; underparts varying from buffy white to light brown, or even reddish brown.

The Common Woodchuck feeds on green plants as do other marmots. It is also a hibernating species, and usually goes to sleep during October, and awakens in February. Woodchucks can be destructive creatures

Figure 134

if they happen to live near a garden. They seldom wander far from their burrows, and run for the burrow at the slightest sign of danger.

Genus SPERMOPHILUS
Ground Squirrels

The ground squirrels have short, rounded ears and short tails which are usually not more, and often much less, than half the total length. Colors are plain gray or brownish gray, often with spots or stripes. The skull has the dorsal profile moderately convex; braincase subglobular, about as broad as long; postorbital process long, slender, and decurved; zygomatic arches heavy; upper incisors slender. All have the ability to dig burrows in the ground, but lack the ability to climb trees. A few have been reported in trees or bushes on rare occasions.

Key to the Species of Ground Squirrels

1a. Upper parts brownish or grayish, or a mixture of both, but never striped or dappled; when gray the hairs on the tail are longer than on the back, and the tail is less than ½ the length of head and body 2

1b. Upper parts striped, or dappled, or red, or gray and dappled; when red or gray the hairs on tail are no longer than on the back, and the tail is more than ½ the length of head and body............9

2a. Upper parts not spotted.......................................3

2b. Upper parts spotted or mottled...............................7

3a. Hind foot more than 39 mm. 4

3b. Hind foot less than 39 mm. 6

4a. Underside of tail gray. Fig. 135. *Spermophilus armatus.* **Uinta Ground Squirrel.**

Total length 280 to 303 mm.; tail 63 to 81 mm.; hind foot 42 to 45 mm.; ear 10 to 12 mm.; color of upper parts light brown to buff; head cinnamon brown sprinkled with gray; underparts buff to buffy white. This ground squirrel lives at rather high altitudes, especially in wooded regions.

Figure 135

4b. Underside of tail buffy or reddish 5

5a. Underside of tail buffy; tail usually more than 75 mm.; *Spermophilus richardsoni.* **Fig. 136. Richardson's Ground Squirrel.**

Total length 277 to 306 mm.; tail 65 to 83 mm.; hind foot 43 to 47 mm.; color of upper parts uniform buff or drab; underparts deep buff. In worn pelage, as at the end of the summer, the upper parts are a faded gray color.

The Richardson Ground Squirrel inhabits the grassy plains and the arid and semiarid regions of the Great Basin states. In some places these squirrels are so tame that they will eat from one's hand. This may be due to a lack of experience with man, for they live in some of the most desolate regions of America.

Figure 136

5b. Underside of tail reddish; tail usually less than 75 mm. Fig. 137.
Spermophilus beldingi. **Belding's Ground Squirrel.**

Total length 268 to 295 mm.; tail 60 to 75 mm.; hind foot 42 to 47 mm.; color of upper parts smoke gray mixed with buff, the middle of the back somewhat darker and rather brownish; underparts dull white or almost buff.

The Belding Squirrels live in wilderness areas of the Great Basin states. They prefer grass for food, but other plants will do. They will sit straight up in their burrows and watch an intruder, then with a shrill whistle they dart down the burrow.

Figure 137

6a. Size about 246 (222 to 271 mm.); tail about 60 (46 to 72 mm.); upper parts dappled. Fig. 138. *Spermophilus idahoensis.* **Idaho Ground Squirrel.**

Hind foot 33 to 38 mm.; color of upper parts pale smoke gray, sometimes faintly shaded with buff, and a slight indication of mottling or dappling; underparts grayish white faintly washed with buff.

Figure 138

6b. Size about 210 (167 to 238 mm.); tail about 45 (32 to 61 mm.); upper parts plain gray. Fig. 139. *Spemophilus townsendii.* **Townsend's Ground Squirrel.**

Hind foot 29 to 37 mm.; color of upper parts plain smoke gray, shaded with buff or cinnamon; tail cinnamon drab or clay color; cinnamon patch on front of face; underparts creamy white washed with buff.

Figure 139

The Townsend Squirrel inhabits open country, where it makes its burrows among the sage brush. Little damage is done by these squirrels unless they happen to live near a garden or irrigated field.

7a. Hind foot, 48 to 58 mm.; total length 325 to 495 mm.............8

7b. Hind foot 34 to 38 mm.; total length 185 to 245 mm. Fig. 140.
Spermophilus washingtoni. Washington Ground Squirrel.

Tail 32 to 65 mm.; color of upper parts pale smoke gray faintly washed with buff, with the entire back covered with small square grayish white spots; cinnamon patch on nose; underparts grayish white washed with pale buff. This squirrel lives among the sage brush or bunch grass hills.

Figure 140

8a. Dorsal spots white; total length about 450 (420 to 495 mm.). Spermophilus parryii. Parry's Ground Squirrel.
Tail 115 to 136 mm.; hind foot 63 to 68 mm.; color of upper parts brown mixed with gray and flecked with large irregular white spots; head tawny or reddish brown; tail reddish below; underparts buff to grayish white.

This large ground squirrel lives in the barren regions of the far north, where it must spend most of the year in hibernation. If one were to travel the Alcan Highway he would enter the territory of this interesting squirrel.

8b. Dorsal spots buff; total length about 360 (340 to 410 mm.). Fig. 141.
Spermophilus columbianus. Columbian Ground Squirrel.

Tail 80 to 120 mm.; hind foot 48 to 58 mm.; color of upper parts grizzled brownish gray, dorsal spots very small; face, feet, and legs brown or even reddish; underparts clay color to cinnamon buff. These squirrels are found in both the valleys and the high mountain meadows.

Figure 141

9a. Dorsal stripes present 10

9b. Dorsal stripes absent 15

10a. Only two stripes present, these white........................11

10b. More than two stripes present................................13

11a. Underside of tail white down the center.....................12

11b. Underside of tail not white down the center. Fig. 142. *Ammospermophilus harrisii.* Gray-tailed Antelope Squirrel.

Figure 142

Total length 223 to 250 mm.; tail 74 to 94 mm.; hind foot 38 to 42 mm.; color of upper parts mouse gray in winter, the hairs having white tips; a narrow white stripe extends from shoulders to rump on each side of the back; upper parts light cinnamon in summer; underparts white or creamy white.

This charming little squirrel (along with the next two species below) is one of the most interesting in the country. It is mistaken for a chipmunk by some, but the two stripes should be sufficient to show the true identity to any careful observer. This squirrel lives in the deserts of the southwestern states, where it is a familiar animal.

12a. Tail hairs with two black bands. Fig. 143. *Ammospermophilus nelsoni.* San Joaquin Antelope Squirrel.

Figure 143

Total length 218 to 240 mm.; tail 63 to 79 mm.; hind foot 40 to 43 mm.; colors similar to *A. harrisii* above, except that the underside of the tail is white.

12b. Tail hairs with one black band. Fig. 144. *Ammospermophilus leucurus.* White-tailed Antelope Squirrel.

Figure 144

Total length 194 to 238 mm.; tail 54 to 79 mm.; hind foot 37 to 43 mm.; two conspicuous white stripes down the back; underside of tail white edged with a narrow band of black.

This beautiful little squirrel, as well as the other two species above, carries its tail up over its back, hence the name Antelope Squirrel. Some people mistake this animal for a chipmunk, but the absence of stripes on the head, and the smaller number of dorsal stripes will distinguish it at a glance.

13a. Two light stripes and four dark stripes (one light and two dark on each side of body) 14

13b. Six light stripes and five dark stripes, as well as several indistinct stripes outside of these eleven. Fig. 145. *Spermophilus tridecemlineatus*. Thirteen-lined Ground Squirrel.

Total length 174 to 297 mm.; tail 69 to 132 mm.; hind foot 27 to 41 mm.; upper parts a series of dark and light stripes; dark stripes brownish or blackish; light stripes whitish; dark stripes with a series of light spots down the center; underparts tawny brown.

This common ground squirrel lives usually in prairies, making its burrows in colonies. It feeds on a wide variety of plants, doing considerable damage in some places. However it eats a large quantity of harmful insects like grasshoppers, grubs, and wireworms.

Figure 145

14a. Hind foot 43 to 49 mm.; underparts very dark; found in the Cascade Mountains of Washington and southern British Columbia. Fig. 146. *Spermophilus saturatus*. Cascade Mantled Ground Squirrel.

Total length 292 to 323 mm.; tail 106 to 120 mm.; ear 20 to 24 mm.; colors like those of *S. lateralis* (see 14b) except that the colors are darker, underparts buff, middle pair of dark stripes much reduced; outer pair of stripes obscure; skull size larger in all respects except the length of nasal bones, which is about the same as in *S. lateralis*.

Figure 146

14b. Hind foot 35 to 44 mm.; underparts not so dark; not found in the Cascades of Washington and southern British Columbia. Fig. 147. *Spermophilus lateralis*. Golden Mantled Ground Squirrel.

Total length 254 to 292 mm.; tail 65 to 107 mm.; ear 15 to 18 mm.; longitudinal white stripe on each side of back, bordered on each side by a black stripe; mantle of cinnamon brown to reddish brown covers head and shoulders; median area of back gray, buff or cinnamon; underparts whitish. In some subspecies the stripes vary from this description considerably, in that they may be rather dull or even obscure; the mantle varies, and is almost absent except in the breeding season.

Figure 147

These are among our most beautiful ground squirrels, and become so tame in national parks that they feed freely from one's hand, or even climb onto the shoulder to take peanuts from one's fingers. Many call this animal a chipmunk, but the larger size, absence of stripes on the head, the mantle, and the total of only six stripes will easily distinguish it from the true chipmunk. A number of subspecies occur in the mountains of the west.

15a. Upper parts spotted 16

15b. Upper parts not spotted 17

16a. Spots in rows. Fig. 148. *Spermophilus mexicanus*. Rio Grande Ground Squirrel.

Total length 280 to 313 mm.; tail 110 to 126 mm.; hind foot 38 to 43 mm.; color of upper parts light brown with square white spots arranged in linear rows, usually nine in number; cinnamon patch on nose; white ring around eye; underparts pale buff. A resident of low, hot valleys.

Figure 148

16b. Spots not in rows. Fig. 149. *Spermophilus spilosoma*. Spotted Ground Squirrel.

Total length 210 to 253 mm.; tail 55 to 88 mm.; hind foot 28 to 38 mm.; color of upper parts drab, cinnamon drab or smoke gray, more or less spotted with rather square white spots, not arranged in rows; underparts white tinged with pale buff.

This is the common ground squirrel of the southwest. It lives in a variety of habitats, but mainly in typical desert where it runs across the sand with the lizards and horned toads. It eats green plants when it can get them, otherwise it feeds on seeds of desert plants. It goes into hibernation in September or early October, emerging again in February or March.

Figure 149

17a. Tail more than 135 mm.; length of skull more than 50 mm.....18

17b. Tail 55 to 102 mm.; length of skull 34 to 40 mm. 20

18a. Nape and shoulders with dark median area. Fig. 150. *Spermophilus beecheyi*. California Ground Squirrel, or Gray Digger.

Total length 400 to 475 mm.; tail 137 to 198 mm.; hind foot 50 to 63 mm.; ear about 30 mm.; color of upper parts brown flecked with buffy white or cinnamon buff; sides of neck and shoulders whitish extending backward as two diverging stripes to middle of back, leaving a dark triangular area between; underparts buff; tail unusually long for a ground squirrel, and rather bushy, almost as bushy as a tree squirrel.

Figure 150

This animal is almost beautiful if one knew nothing of his habits, but he is so destructive to crops, stored food, grains, and the like that he is quite unpopular. He also carries fleas which harbor the organisms of tularemia and bubonic plague.

18b. Nape and shoulders without a dark median area............19

19a. Ears very short, only 10 to 11 mm. Fig. 151. *Spermophilus franklinii.* **Franklin's Ground Squirrel.**

Total length 381 to 397 mm.; tail 136 to 153 mm.; hind foot 53 to 57 mm.; color of upper parts gray or buffy gray, somewhat grizzled with dark brown hairs; tail grizzled with blackish hairs;

underparts buff or even whitish; the whole color pattern often appears to be somewhat spotted, although no definite spots can be found.

This large squirrel looks somewhat like a gray squirrel, but a careful observer will soon distinguish it. It inhabits grassy areas where there are bushes or hedges nearby, and moves into new territory when timber is cut off. It is usually not abundant anywhere, but appears to be spreading its range in all directions.

Figure 151

19b. Ears long for a ground squirrel, 18 to 21 mm. Fig. 152. *Spermophilus variegatus.* **Rock Squirrel.**

Total length 430 to 540 mm.; tail 174 to 263 mm.; hind foot 56 to 67 mm.; color of upper parts varying from grayish white mixed with cinnamon to dark blackish brown; in some subspecies the head and foreback is black, in others almost the entire back is blackish; underparts dark buff.

The Rock Squirrels live in rocky places in the southwest, or in brushy areas in foothills of mountains, or in brushy canyons even where the elevation is low and the temperature hot. They climb into bushes for nuts and seeds, and live much more like tree squirrels in this respect. They feed on a wide variety of plant foods, not confining their diet to green plants as do most of the ground squirrels.

Figure 152

20a. Underside of tail white. Fig. 153. *Spermophilus mohavensis.* Mohave Ground Squirrel.

Total length 210 to 230 mm.; tail 57 to 72 mm.; hind foot 32 to 38 mm.; color of upper parts light drab to light gray; underparts creamy white; tail round.

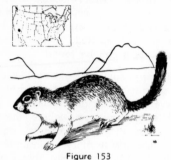

This little squirrel is similar in habits to the Round-tailed Ground Squirrel (20b). It lives in the hot dry Mohave desert where it burrows into the sand, feeds upon the few desert plants, and does the best it can under the circumstances.

Figure 153

20b. Underside of tail drab or buff. Fig. 154. *Spermophilus tereticaudus.* Round-tailed Ground Squirrel.

Total length 204 to 266 mm.; tail 60 to 102 mm.; hind foot 32 to 40 mm.; ears so small they are a mere rim; color of upper parts cinnamon to light drab; underparts buff, tail about the same color all around; tail very round and not densely furred.

These peculiar fellows live in the midst of the hottest, driest desert places they could find. When you stop your car you cannot see them, for they sit motionless, and they match the sand perfectly. How they survive the rigors of the desert is a mystery. Many desert animals, such as these, live all their lives without tasting water, evidently obtaining the moisture they need from plants.

Figure 154

Genus TAMIAS
Eastern Chipmunk

The eastern chipmunk is small compared with other squirrels, with a tail that is not very long, and certainly not bushy. They have large internal cheek pouches in which they store away seeds to carry to their den. They generally nest in the ground, or in an old stump, where they spend much of the winter in a state of hibernation, although they wake up at intervals, run about in the snow, and then go back to sleep again.

Fig. 155. *Tamias striatus*. Eastern Chipmunk.

Total length 215 to 230 mm.; tail 78 to 96 mm.; hind foot 32 to 36 mm.; ear 14 to 16 mm.; color of upper parts reddish brown, with five dark stripes and two light buffy stripes, and a reddish brown rump patch. Stripes present on sides of head as well as on back.

The eastern chipmunk differs from the western chipmunk in that the stripes do not continue to the base of the tail, and there is a reddish rump patch. This charming little animal is fond of a great variety of food, but prefers seeds to almost everything else. He stays on or near the ground most of the time, but is expert at climbing, and thinks nothing of running up the side of a tree in search of seeds. There is only the one species.

Figure 155

Genus EUTAMIAS
Western Chipmunks

The western chipmunk has much more pronounced stripes than does its eastern relative, *Tamias*, for the head stripes are much longer, continuing from the nose to the ears. The back stripes are longer, too, extending from the base of the ears to the base of the tail, or onto the tail. There are four light stripes and five dark stripes, two more stripes than *Tamias*. Its habits are about the same.

Key to the Species of Western Chipmunks

1a. Dorsal stripes (except median one) rather indistinct...........2

1b. Dorsal stripes all distinct...................................3

2a. Postauricular patches large and clearly defined. Fig. 156. *Euta-mias dorsalis*. Cliff Chipmunk.

Total length 222 to 242 mm.; tail 98 to 114 mm.; hind foot 34 to 36 mm.; ear 16 to 18 mm.; color of upper parts pale smoke gray; dorsal stripes not distinct, and never tawny; underparts creamy white. Found in rocky mountainous areas.

Figure 156

2b. Postauricular patches small and not clearly defined. Fig. 157. *Eutamias merriami*. Merriam Chipmunk.

Total length 208 to 277 mm.; tail 95 to 130 mm.; hind foot 33 to 39 mm.; ear 15 to 19 mm.; color of upper parts brown; dark dorsal stripes brown to black; light stripes pale smoke gray to grayish white; underside of tail tawny to pale russet; underparts creamy white.

This is one of the largest of the chipmunks, and may be distinguished from the small chipmunks by its dark colors. It is much like the Townsend Chipmunk, but fortunately the ranges do not overlap.

Figure 157

Note: The Merriam chipmunks north of San Francisco have been named *Eutamias sonomae*, while those south of San Francisco are still *Eutamias merriami*.

3a. Size large, total length often 250 mm. or more; length of skull 37 mm. or more **4**

3b. Size somewhat smaller; length of skull less than 37 mm.........**8**

4a. Postauricular patches large and conspicuous; ears long and narrow, 17 to 20 mm. Fig. 158. *Eutamias quadrimaculatus.* **Long-eared Chipmunk.**

Figure 158

Total length 230 to 250 mm.; tail 98 to 112 mm.; hind foot 35 to 37 mm.; color of upper parts brown mixed with grayish white; rump and thighs grayish; median dorsal stripe brownish black; other dark stripes dark brown; median pair of light stripes grayish white, outer pair creamy white; underparts grayish white; underside of tail tawny to brown, bordered with black and edged with pale gray.

This chipmunk lives in open timber in the Sierras where there are many manzanita bushes and fallen logs. It does not range into the typical Canadian zone but remains in the pines.

4b. Postauricular patches smaller and not conspicuous; ears shorter and broad, about 15 to 18 mm. **5**

5a. Median pair of light dorsal stripes tawny, never white or gray...6

5b. Median pair of light dorsal stripes gray or white..............7

6a. Underparts white. Fig. 159. *Eutamias townsendii.* **Townsend Chipmunk.**

Figure 159

Total length 235 to 263 mm.; tail 96 to 125 mm.; hind foot 34 to 36 mm.; ear 15 to 17½ mm.; color of upper parts brown to tawny brown; dark dorsal stripes black or brownish black; light dorsal stripes tawny to buffy white; tail tawny below, bordered with black and tipped with pale gray; underparts creamy white. Subspecies vary a great deal in color. This is a lowland chipmunk as a rule, but subspecies occur in nearby mountains. A closely related species, *Eutamias sonomae*, occurs in northern California.

6b. Underparts buff. Fig. 160. *Eutamias alleni*. Marin Chipmunk.
Total length 231 to 250 mm.; tail 100 to 113 mm.; hind foot 34 to
37 mm.; ear 15 to 18 mm.; color of upper parts reddish brown mixed
with gray; dark dorsal stripes black; light dorsal stripes tawny
(median pair); outer pair of light stripes dull whitish; underside
of tail tawny bordered with black and tipped with pale buff.

Figure 160

This species, which is very similar to the Townsend Chipmunk, lives on the point of land just
to the north of the Golden Gate of San Francisco.
It inhabits the wooded areas near the coast, and
is separated from other similar chipmunks by the
salt marshes of San Francisco Bay.

7a. General tone of upper parts cinnamon buff. Fig. 161. *Eutamias townsendii cooperi*. Cooper Chipmunk.

Total length 238 to 263 mm.; tail 102 to 120 mm.; hind foot 34 to
38 mm.; ear 15 to 18 mm.; dark dorsal stripes black or brownish

Figure 161

black; light dorsal stripes grayish white, the median pair often buffy. This chipmunk is a subspecies of the Townsend chipmunk, but since it
is so much paler in color it cannot be keyed with
the Townsend chipmunks.

7b. Upper parts tawny or grayish. *Eutamias merriami*. Merriam Chipmunk.
This chipmunk varies a great deal in the colors and sizes of the
subspecies; while most of them are dark in color, some are lighter,
so I include it here. For the complete description see 2b above,
keeping in mind that the colors may be lighter than the general description indicates.

11a. **Tail rather bushy; ear longer, 12 to 14 mm.; interorbital region 6.7 to 10.1 mm. Fig. 162.** *Eutamias alpinus.* **Alpine Chipmunk.**

Figure 162

Total length 176 to 195 mm.; tail 70 to 85 mm.; hind foot 28 to 31 mm.; color of upper parts more tawny than in *E minimus;* dark dorsal stripes tawny; median pair of light dorsal stripes smoke gray; outer pair creamy white; underside of tail clay color washed with pinkish cinnamon; underparts creamy white.

Superficially this chipmunk resembles the species *minimus,* but its broad interorbital region makes it distinct from all other chipmunks. It occurs only in the high Sierras of California.

11b. **Tail not bushy; ear shorter, 10 to 11½ mm.; interorbital region 6 to 7 mm. Fig. 163.** *Eutamias minimus.* **Least Chipmunk.**

Total length 167 to 200 mm.; tail 74 to 90 mm.; hind foot 26 to 30 mm.; color of upper parts rather grayish; median dorsal dark stripes black, outer pair dark brown; light dorsal stripes grayish white, the median stripes tinged with buff; underside of tail brown to grayish; underparts creamy white.

Figure 163

This little chipmunk occurs in many of the desert areas of the western states, and into moderately high mountains, especially arid ranges. In many places it is the only chipmunk not found in coniferous forests.

12a. **Skull longer, usually 35 to 36.8 mm. (sometimes as small as 34.5 mm.); dark dorsal stripes black, margined with tawny** 13

12b. **Skull shorter, usually less than 35 mm. (31.3-35.8 mm., but rarely over 34.5 mm.); dark dorsal stripes black, mixed with tawny** 15

13a. **Found at Charleston Peak, Nevada; outer pair of dorsal stripes clear white; submalar stripe brownish. Fig. 164.** *Eutamias palmeri.* **Palmer Chipmunk.**

Figure 164

Total length 210 to 223 mm.; tail 86 to 101 mm.; hind foot 32 to 34 mm.; ear 13½ to 15½ mm.; color of upper parts tawny brown; dark dorsal stripes dark brown, the median one blackish brown bordered with brown; light stripes: median pair light smoke gray, outer pair clear white; underside of tail tawny bordered with black and edged with buff; underparts creamy white.

This is another isolated species, occurring only in the Canadian Life Zone of Charleston Peak, Nevada. It is most nearly related to *E. quadrivittatus.*

13b. Not found at Charleston Peak, Nevada......................**14**

14a. Found in the San Jacinto, San Bernardino and Piute Mountains, California; outer pair of dorsal stripes buffy white; submalar stripe blackish. Fig. 165 . Eutamias speciosus. Sierra Chipmunk.

Figure 165

Total length 203 to 231 mm.; tail 84 to 100 mm.; hind foot 33 to 35 mm.; ear 14½ to 16 mm.; color of upper parts pale tawny; median dorsal stripe blackish with brown borders; outer dark stripes brownish black, median light stripes grayish white; outer light stripes creamy white; underside of tail tawny, bordered with black and edged with buff; underparts creamy white.

14b. Found in several mountain ranges in California and Nevada where 14a and 13a do not occur; outer pair of dorsal stripes creamy white. Fig. 166. Eutamias quadrivittatus. Colorado Chipmunk. (Common names vary with the subspecies, so this name does not apply to all of them).

Figure 166

Total length 216 to 230 mm.; tail 93 to 104 mm.; hind foot 33 to 35 mm.; ear 14 to 16 mm.; color of upper parts brownish gray; dark dorsal stripes black, margined with tawny, the outer pair sometimes mainly tawny; light dorsal stripes grayish white, the outer pair usually creamy white; underside of tail tawny, bordered with black and edged with buff, underparts creamy white.

The chipmunks in this species vary as much as do others. They live in the mountains of the states of Colorado, Utah, Arizona, Nevada, California, and New Mexico, but each range of mountains is isolated from the others so that each range has its own subspecies of chipmunk.

15a. Dark dorsal stripes black, or more blackish than reddish, never gray; submalar dark stripe complete anteriorly; dorsal face of skull rounded. Fig. 167. *Eutamias amoenus*. Common Western Chipmunk.

Total length 181 to 230 mm.; tail 78 to 107 mm.; hind foot 30 to 34 mm.; ear 12½ to 15 mm.; color of upper parts varying from grayish brown to tawny; dark dorsal stripes black or brownish black; median pair of light stripes smoke gray, outer pair creamy white; underside of tail gray, buff, or brown; underparts ceamy white, often washed with buff.

Figure 167

This chipmunk is the most familiar one in the west, and occurs in most of the timbered areas of those states, except along the coastal areas of the three Pacific states; it is also absent from the southwestern states.

15b. Dark dorsal stripes, except median one, either reddish, grayish, or almost obsolete; submalar dark stripe obsolete anteriorly; dorsal face of skull flattened. Fig. 168. *Eutamias panamintinus*. Panamint Chipmunk.

Total length 198 to 220 mm.; tail 85 to 102 mm.; hind foot 31 to 32½ mm.; ear 13½ to 16 mm.; color of upper parts tawny; dark dorsal stripes brown; median pair of light stripes grayish white, outer pair creamy white; rump gray.

Figure 168

This chipmunk is rather characteristic with its tawny upper parts and gray rump. It occurs only in the high mountain ranges of the southern Sierra Nevadas and the western Nevada ranges. It lives mainly in the pinon pine areas.

16a. Size smaller, total length usually about 200 mm., length of skull usually less than 33.4 mm.; color decidedly grayish. *Eutamias minimus* (see 11b).

16b. Size larger, total length usually more than 210 mm., length of skull usually more than 33.4, often more than 34.5.; color not grayish but buffy, tawny or even reddish brown............17

17a. Shoulders washed with gray. Fig. 169. *Eutamias cinereicollis.* Gray-collared Chipmunk.

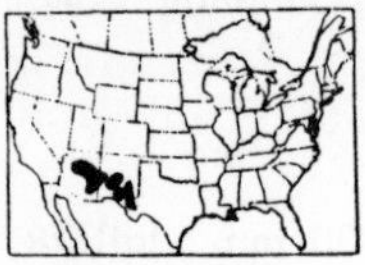

Total length 212 to 242 mm.; tail 95 to 109 mm.; hind foot 34 to 36 mm.; ear 14 to 16 mm.; color of upper parts less tawny and more grayish; dark dorsal stripes black, bordered with brown; light dorsal stripes pale smoke gray; underside of tail tawny bordered with black and edged with buff; underparts creamy white.

Figure 169

This chipmunk inhabits the open timber in the rather high plateaus.

17b. Shoulders not washed with gray.............................18

18a. Dorsal stripes blackish 19

18b. Dorsal stripes brown. Fig. 170. *Eutamias umbrinus.* Uinta Chipmunk.

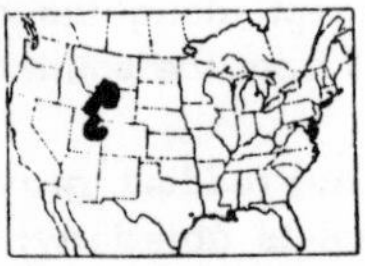

Total length 216 to 240 mm.; tail 92 to 113 mm.; hind foot 31 to 33 mm.; ear 13 to 15 mm.; color of upper parts reddish brown mixed with some gray; median dorsal stripe black edged with brown; outer dark stripes brown; light stripes white, the median pair tinged with brown; underside of tail tawny to brown, bordered with black and edged with buff; underparts creamy white. This chipmunk mhabits the timbered areas of the Uinta Mountains in Utah.

Figure 170

19a. Tail tawny beneath. *Eutamias quadrivittatus.* Colorado Chipmunk. (See 14b above for description.)

19b. Tail brown to reddish brown beneath. Fig. 171. *Eutamias adsitus.* Beaver Mountain Chipmunk.

Total length 209 to 229 mm.; tail 86 to 99 mm.; hind foot 31 to 33 mm.; ear 12 to 14 mm.; color of upper parts reddish brown mixed with gray; dark dorsal stripes black; median pair of light stripes grayish white, the outer pair pure white; underside of tail brown or tawny; underparts white. This is the common chipmunk of the Beaver Mountains and of the Kaibab Plateau of the north side of the Grand Canyon.

Figure 171

20a. Size smallest, total length usually about 200 mm., length of skull usually less than 33.4 mm.; color decidedly grayish. *E. minimus* (see 11b).

20b. Size larger, total length 181 to 250 mm., but usually more than 200 mm.; length of skull 31.3 to 36.8 mm., but usually more than 33.4 mm.; color not grayish, but buff, tawny, or reddish brown..21

21a. Underside of tail tawny to amber brown, but not reddish or pinkish; size averages larger, total length 204 to 248 mm., length of skull averages 33.4 to 36.8 mm. 22

21b. Underside of tail not tawny but cinnamon brown to pinkish cinnamon; size averages smaller, total length 181 to 236 mm.; length of skull averages 31.3 to 35.6 mm. (Note: these four chipmunks are very difficult to distinguish from each other; in some cases the range will help, but in others it will be almost impossible for a beginner to find the correct species for sure). Fig. 167. *Eutamias amoenus.* (See 15a).

22a. Color of head drab or grayish, paler than in 22b..............23

22b. Color of head tawny to cinnamon brown. Fig. 172. *Eutamias ruficaudus.* Rufous-tailed Chipmunk.
Total length 223 to 248 mm.; tail 101 to 121 mm.; hind foot 32 to 36 mm.; ear 13 to 15 mm.; color of upper parts tawny to reddish

brown; dark dorsal stripes black or brownish black; median pair of light stripes grayish white mixed with tawny; outer pair of light stripes creamy white; underside of tail amber brown bordered with black and edged with cinnamon; underparts creamy white washed with pale buff.

Figure 172

This chipmunk is decidedly reddish, especially the underside of the tail; its large size will also help to distinguish it from the other chipmunks of the same area.

23a. Dorsal stripes blackish. Fig. 166. *E. quadrivittatus* (see 14b).

23b. Dorsal stripes brownish. Fig. 170. *E. umbrinus* (see 18b).

Genus GLAUCOMYS
Flying Squirrels

Flying squirrels are easy to recognize with their soft silky fur, long flattened tails, big brown eyes, and flying membrane, or patagium, along the sides of the body. Yet, we seldom see them for they remain hidden in their nests during the day, coming out to hunt for food only at night. The beautiful animals are often abundant in areas where most people scarcely know they exist. Only two species occur throughout the country and because of the great differences in size, it will be no trouble to tell them apart.

Key to the Species of Flying Squirrels

1a. Size small, total length 211 to 253 mm.; tail 81 to 115 mm.; color of underparts pure white. Fig. 173. *Glaucomys volans*. Eastern Flying Squirrel.
Hind foot 28 to 33 mm.; color of upper parts drab to brown.

This flying squirrel lives throughout the wooded areas of the eastern states where it nests in hollow trees. It really does not fly, but merely glides from one tree to the next by climbing to the top of the tree, then launching forth through the air in a downward direction with its legs and tail spread as far as possible, landing at the base of another tree often some distance away. It feeds on nuts, berries, and a host of animal food, being very fond of meat. It destroys numbers of birds nests.

Figure 173

1b. Size much larger, total length 263 to 365 mm.; tail 128 to 180 mm.; color of underparts white, buffy white, or buff. Fig. 174. *Glaucomys sabrinus*. Northern Flying Squirrel.

Hind foot 34 to 45 mm.; color of upper parts varying all the way from drab through all shades of brown to dark reddish brown, the redder colors occuring in coast subspecies.

The habits of this squirrel are similar to those of the eastern species. All flying squirrels have bodies much smaller actually than they appear to have. This accounts for their agility of flight from one tree to another. Flying squirrels usually bring forth young in April, but a second brood may appear in August or September.

Figure 174

Genus CYNOMYS

Prairie Dogs

These plump rodents are characteristic animals of the prairie regions of the midwestern states. Their bodies are very stout, tails short and flat; ears small; cheek pouches shallow. They nest in colonies where they excavate rather extensive burrows, carrying out the soil and piling up a mound at the entrance of the burrow. One can generally see a number of them standing upright at the entrances of their burrows. When one approaches the colony, the prairie dogs all flop down into the burrows—coming up cautiously a few moments later.

Key to Species of Prairie Dogs

1a. Tail tipped with black. Fig. 175. *Cynomys ludovicianus.* Black-tailed Prairie Dog.

Total length 350 to 415 mm.; tail 75 to 100 mm.; hind foot 57 to 65 mm.; color of upper parts cinnamon buff; underparts white to buffy white.

Colors vary a great deal according to the season, but they are usually buff to some extent, at least. Prairie dogs usually hibernate, but do not remain in hibernation for as long periods as do ground squirrels. Green plants form the food for these animals, and they consume great quantities of them.

Figure 175

1b. Tail tipped and bordered with white..........................2

2a. Terminal half of tail white without dark center................3

2b. Terminal half of tail bordered and tipped with white, with a gray center. Fig. 176. *Cynomys gunnisoni.* Gunnison Prairie Dog.

Total length 309 to 373 mm.; tail 39 to 68 mm.; hind foot 52 to 62 mm.; color of upper parts drab to yellowish drab; underparts buff.

Figure 176

3a. Color in summer buff or grayish, never reddish; size larger: total length 340 to 370 mm.; tail 44 to 60 mm.; hind foot 60 to 65 mm. Fig. 177. *Cynomys leucurus*. White-tailed Prairie Dog.

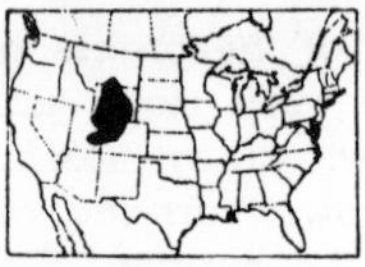

Color of upper parts buffy to grayish; underparts clear buff.

This prairie dog, and the next species below, look more like ground squirrels than does *Cynomys ludovicianus.*

Figure 177

3b. Color in summer reddish to rich cinnamon, never buff or gray; size somewhat smaller; total length 305 to 360 mm.; tail 30 to 57 mm.; hind foot 55 to 61 mm. Fig. 178. *Cynomys parvidens*. Utah Prairie Dog.

The color is the best distinguishing character between these two prairie dogs, yet, since the ranges of the two do not overlap there is little chance that they should be confused.

Figure 178

Family GEOMYIDAE
Pocket Gophers

Pocket gophers occur in most of the areas of the United States, although there are some places where they are still absent. They are loose skinned, thick-bodied rodents with large external cheek pouches, bare tails, very long incisors, and extremely short external ears. Their eyes are small and beady. The fore feet are enlarged with elongated claws; the hind feet are smaller, with shorter claws. They make extensive burrows under the surface of the ground where they tunnel into every situation that feels good to them, pushing up dirt into crescent-shaped mounds at the exit hole. Main tunnels are often a foot or so below the ground; surface branches are used for getting rid of soil. Their food consists mainly of roots and tender shoots of plants, but they often come to the surface for green leaves and stems of plants. They emerge mainly at night, but one may also see them during the day when they come up to push out dirt. Many believe that dirt is carried out in the cheek pouches, but this is erroneous. The pouches are used to store grass and other plants to carry back to their burrows.

Key to Genera and Species of Pocket Gophers

Note: The species of *Thomomys* are difficult to identify. This key is not intended to be anything more than an attempt to separate the species. It is best to send specimens to an expert for correct identification. There is so much variation in pocket gophers that even authorities on the subject have trouble with them, and the system of classification is always being changed. Use the key only for tentative identification.

1a. Outer surface of upper incisors grooved........................2

1b. Outer surface of upper incisors not grooved. *Thomomys*........9

2a. Upper incisors with two grooves in each tooth. Geomys........3

2b. Upper incisors with only one groove. Fig. 179. *Cratogeomys castanops*. Chestnut-faced Pocket Gopher.

Total length 260 to 295 mm.; tail about 65 to 95 mm.; hind foot about 37 mm.; color of upper parts yellowish brown mixed with black-tipped hairs; tip of tail black; underparts buff.

Figure 179

3a. Color dark brown to dark reddish brown......................4

3b. Color pale brown to pale drab; found in southern Texas and southern New Mexico..8

4a. Found in Florida, Alabama and Georgia......................5

4b. Found in the upper Mississippi Valley. Fig. 180. *Geomys bursarius*. Plains Pocket Gopher.

Total length 270 to 325 mm.; tail 75 to 95 mm.; hind foot 34 to 37 mm.; color of upper parts dark reddish brown; underparts lighter brown.

This is the common pocket gopher of the northern part of the Mississippi Valley. It is one of the very largest of the pocket gophers, the males often reaching a total length of more than one foot.

Figure 180

5a. Found only on Cumberland Island, Georgia. *Geomys cumberlandius*. Cumberland Island Pocket Gopher.

Total length, 324 mm.; tail, 114 mm.; hind foot, 35.5 mm. Similar to *Geomys pinetis*, except for a longer tail.

As often occurs, a mammal living on an island is found to be a different but closely related species to that found on the nearby mainland.

5b. Not found on Cumberland Island, Georgia.....................6

6a. Size smaller, total length 222 to 276 mm. in males and 232 to 250 mm. in females; hind foot small, 30 to 34 mm. Found only in one small area, 7 miles NW of Savannah, Georgia. *Geomys fontenelus*. Sherman's Pocket Gopher.

Here is a case where a distinct species is found in an unusually small area. Perhaps further study will reveal more about this situation.

6b. Size larger; found in other regions...........................7

7a. Nasal bones greatly constricted near the center in the form of an hourglass; palate narrower; color dark brown to blackish; found in southern Alabama, southern Georgia, and northern Florida. *Geomys pinetis*. Southeastern Pocket Gopher.

Total length, 250 to 305 mm. (males); 229 to 335 mm. (females); tail, 76 to 96 mm.; hind foot, 31 to 37 mm.

This is the common pocket gopher of this area.

7b. Nasal bones scarcely constricted, and not shaped like an hourglass; palate wider; color darker than in 7a; found only near St. Marys, Camden County, Georgia. *Geomys colonus*. Colonial Pocket Gopher.

Total length (males) 280 to 288 mm.; tail, 89 to 100 mm.; hind foot, 34 to 36 mm.

This is another case where a distinct species occurs in an extremely restricted area, yet on the mainland near other species of gophers. More study is needed here, also.

8a. Size large, total length 275 to 310 mm.; hind foot 34 to 38 mm. Fig. 181. *Geomys personatus*. Padre Island Pocket Gopher.

Tail 83 to 110 mm.; upper parts pale brown; underparts whitish; found from Padre Island south to Carrizo on the Rio Grande, Texas.

Figure 181

8b. Size medium, 250 to 265 mm.; hind foot 33 mm. *Geomys arenarius.*
Fig. 182. Sand Pocket Gopher or Desert Pocket Gopher.

Tail about 88 mm.; upper parts pale buffy brown; underparts whitish.

Figure 182

9a. Rostrum deep and evenly sloping in front of upper molars....10
9b. Rostrum slender; abruptly arched in front of upper molars....15

10a. Pterygoids concave on inner surface, and convex on outer; size very large Fig. 183, about 300 mm.; mammae in 4 pairs. Fig. 184. *Thomomys bulbivorus.* Camas Pocket Gopher.

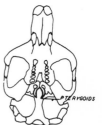

Figure 183

Total length 275 to 302 mm.; tail 81 to 99 mm.; hind foot 39 to 43 mm.; color of upper parts dark sooty brown; underparts nearly as dark.

This is the largest of the western pocket gophers, and its tunnels are very large in diameter. When these gophers get into the garden, residents of western Oregon must abandon the garden or get rid of the gophers.

Figure 184

10b. Pterygoids flat and straight; size much smaller.................11

11a. Skull short and wide. **Fig. 185.** *Thomomys umbrinus.* Southern Pocket Gopher.

Total length 167 to 260 mm.; tail 51 to 94 mm.; hind foot 25 to 33 mm.; ear only 5 to 8 mm.; color of upper parts gray or buff; underparts whitish; mammae in 4 pairs; skull with sphenorbital fissure.

This is the most common gopher in the southwestern states, and can be distinguished by its pale colors.

Figure 185

A closely related species, *Thomomys baileyi,* occurs in southern Arizona, southern New Mexico and southern Texas, in widely isolated areas. This species will likely be placed in *T. umbrinus* one of these days.

11b. Skull longer and more slender...............................12

12a. Color very dark. Fig. 186. *Thomomys alpinus. Alpine Pocket Gopher.*

Figure 186

Total length 222 mm.; tail 61 mm.; hind foot 30 mm.; color of upper parts dark gray or brownish gray almost black in the center of the back; underparts paler. Occurs in the high mountains of California.

12b. Color pale 13

13a. Color buffy or yellowish, gray or black......................14

13b. Color tawny. Fig. 187. *Thomomys fulvus.* Fulvous Pocket Gopher.

Figure 187

Total length 200 to 244 mm.; tail 63 to 76 mm.; hind foot 26 to 31 mm.; color of upper parts dark tawny to light chestnut; underparts pale tawny.

14a. Color buffy or yellowish. Fig. 188. *Thomomys perpallidus.* Palm Springs Pocket Gopher.

Figure 188

Total length 215 to 272 mm.; tail 66 to 100 mm.; hind foot 27 to 35 mm.; color of upper parts buff to yellowish; underparts whitish.

14b. Color gray or blackish. Fig. 189. *Thomomys townsendii.* Townsend Pocket Gopher.

Figure 189

Total length 221 to 305 mm.; tail 61 to 100 mm.; hind foot 28 to 38 mm.; color dark buffy gray to sooty gray; underparts buffy.

15a. Mammary glands in 6 pairs. **Fig. 190.** *Thomomys talpoides.* Common names vary with the subspecies.

Total length 203 to 240 mm.; tail 34 to 72 mm.; hind foot 26 to 31 mm.; color of upper parts grayish brown to brownish gray, often even reddish brown.

This is the common pocket gopher in the northwestern and the northern Rocky Mountain states. Since the habitats available to pocket gophers are so varied in these states, and since there are so many mountain ranges, large rivers, and other natural barriers, a great many sub-

Figure 190

species occur. In some cases there seems to be a different subspecies over every little hill and down in every small river valley.

15b. Mammae in 4 or 5 pairs....................................16

16a. Mammae in 5 pairs. **Fig. 191.** *Thomomys fossor.* **Colorado Pocket Gopher.**

Figure 191

Total length about 220 to 240 mm.; tail about 60 to 70 mm.; hind foot 29 to 34 mm.; color of upper parts dull to rich brown; underparts buff.

16b. Mammae in 4 pairs. **Fig. 192.** *Thomomys monticola.* **Mountain Pocket Gopher.**

Figure 192

Total length 202 to 214 mm.; tail 55 to 76 mm.; hind foot 26 to 29 mm.; color of upper parts dark brown to grayish brown; underparts buff.

Family HETEROMYIDAE

Pocket Mice

This family may be recognized at once by the presence of fur-lined external cheek pouches, by the low, wide, flat skulls with very large auditory bullae, and by the long tails. Pocket mice are found mainly in arid and semi-arid regions.

Key to the Genera of Pocket Mice

1a. Soles of hind feet naked; fore feet not very much smaller than hind feet; greatest width of skull less than distance between tip of nose and posterior end of eye; animals do not jump about on their hind feet. Genus *Perognathus*, page 114.

1b. Soles of hind feet haired or furred; fore feet much smaller than hind feet; greatest width of skull more than the distance between tip of nose and posterior end of eye; animals commonly jump about on hind feet 2

2a. Tail with terminal tuft; size very large for a mouse, with the hind foot more than 32 mm. *Dipodomys*, page 120.

2b. Tail without terminal tuft; size average or even small for a mouse, with the hind foot less than 32 mm. *Microdipodops*, page 126.

Genus PEROGNATHUS

Pocket Mice

Pocket mice are medium-sized mice with grizzled grayish and buffy colors, with pure white underparts, and fairly long tails. The cheek pouches are very evident, and the head is rather wide. They make their burrows in dry bushy areas, or along the roadside, where they can find plenty of weed seeds, or grain to eat. They do some damage to crops if they invade grain fields, but they eat weed seeds most of the time, storing them in their burrows. Some mammalogists believe that these mice hibernate during very cold weather. Pocket mice occur almost entirely in the western states.

Key to Species of Pocket Mice

1a. Size medium or large; fur harsh, often with spiny bristles on rump; mastoids small, not projecting beyond plane of occiput; auditory bullae separated by nearly full width of basisphenoid..........2

1b. Size medium or small; fur soft with no spines; mastoids greatly developed, projecting beyond plane of occiput; auditory bullae meeting or nearly meeting anteriorly 10

2a. Rump with distinct spines or bristles or both 6

2b. Rump without spines or bristles 3

3a. Tail shorter than head and body, and not crested. Fig. 193. *Perognathus hispidus.* Hispid Pocket Mouse.

Perognathus hispidus. Hispid Pocket Mouse.

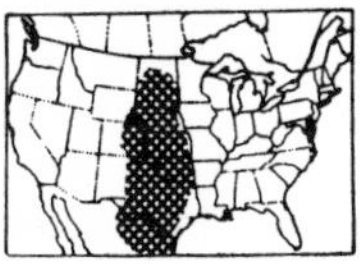

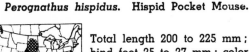

Total length 200 to 225 mm; tail 100 to 108 mm.; hind foot 25 to 27 mm.; color of upper parts buffy gray mixed with blackish hairs; underparts white.

Figure 193

3b. Tail longer than head and body, and crested 4

4a. Total length more than 200 mm. 5

4b. Total length less than 200 mm. *Perognathus penicillatus* (see 5b.)

5a. Total length usually more than 210 mm.; color of upper parts grayish. Fig. 194. *Perognathus baileyi.* Bailey's Pocket Mouse.

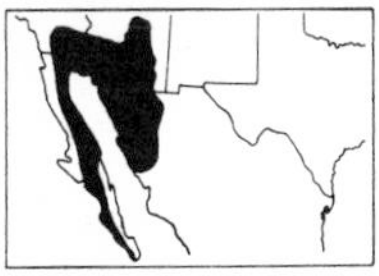

Total length 201 to 230 mm.; tail 110 to 125 mm.; hind foot 26 to 28 mm.; color of upper parts grayish, washed with yellowish or buff; underparts white or nearly so; tail long with a prominent crest at the tip, buff above and white below.

Figure 194

5b. Total length usually less than 210 mm., usually around 200 mm.; color of upper parts yellowish brown. Fig. 195. *Perognathus penicillatus.* Desert Pocket Mouse.

Total length about 200 mm.; tail about 110 mm.; hind foot about 25 mm.; color buff with blackish hairs mixed through the buff; underparts white.

Figure 195

6a. Strong spines on rump and flanks; lateral line faint or lacking. Fig. 196. *Perognathus spinatus.* Spiny Pocket Mouse.

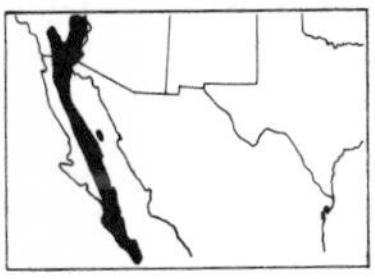

Total length 164 to 225 mm.; tail 89 to 128 mm.; hind foot 20 to 28 mm.; upper parts brownish to buffy yellow; underparts white or buffy white; tail long and crested, brownish above, white below.

Figure 196

6b. Spines present on rump only, and not unusually stiff 7

7a. Ear less than 9 mm. 8

7b. Ear more than 9 mm.; size large; tail long. Fig. 197. *Perognathus californicus.* California Pocket Mouse.

Figure 197

Total length 190 to 235 mm.; tail 103 to 143 mm.; hind foot 24 to 29 mm.; upper parts brownish gray with buffy flecks; underparts yellowish white; tail crested, darker above, whitish below.

8a. Spines on rump fairly stiff; size fairly large; tail fairly long. Fig. 198. *Perognathus fallax.* San Diego Pocket Mouse.

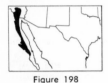

Figure 198

Total length 176 to 200 mm.; tail 88 to 118 mm.; hind foot about 23 mm.; upper parts deep brown, somewhat blackish over rump; underparts white or whitish; buffy lateral line; tail crested, dark above and light below.

8b. Spines on rump weakly developed.............................9

9a. Total length more than 180 mm.; size moderately large. Fig. 199. *Perognathus nelsoni.* Nelson's Pocket Mouse.

Figure 199

Total length 182 to 193 mm.; tail 104 to 117 mm.; hind foot 22 to 23 mm.; upper parts dark brown; underparts white or whitish; prominent lateral line is light buff; tail with a prominent crest; tail blackish above, white below.

9b. Total length 180 mm. or less. Fig. 200. *Perognathus intermedius.* Rock Pocket Mouse.

Figure 200

Total length about 180 mm.; tail about 100 to 105 mm.; hind foot 22 to 23 mm.; ear 7 mm.; color of upper parts drab, with a blackish wash on back and rump; underparts white.

This pocket mouse prefers to live in the desert-like regions of the Southwest from the Grand Canyon country southward and eastward into New Mexico, then still farther south into Mexico. It usually occurs in the lowlands rather than into the hills, but it does occur in the desert ranges as well.

10a. Hind foot more than 20 mm.................................11

10b. Hind foot 20 mm. or less...............................14

11a. Ears covered with white hairs; tail pale in color...............12

11b. Ears not covered with white hairs; tail dark in color..........13

12a. Color of upper parts dark, olive buff to wood brown; underparts white. Fig. 201. *Perognathus alticola*. White-eared Pocket Mouse.

Figure 201

Total length 160 to 181 mm.; tail 72 to 97 mm.; hind foot 21 to 23 mm.; lateral line faint; tail bicolor, the same color as back above, and white below with a blackish tip.

12b. Color of upper parts buff or creamy buff; underparts white. Fig. 202. *Perognathus xanthonotus*. Yellow-eared Pocket Mouse.

Figure 202

Total length about 170 mm.; tail about 85 mm.; hind foot about 23 mm.; tail with a crest; tail light cream colored above and white below.

13a. Soles of hind feet without hairs; tail crested and tufted. Fig. 203 *Perognathus formosus*. Long-tailed Pocket Mouse.

Figure 203

Total length about 190 mm.; tail about 106 mm.; hind foot 24 mm.; color of upper parts grizzled buff; underparts white.

13b. Soles of hind feet with a few hairs; tail not crested nor tufted. Fig. 204. *Perognathus parvus*. Great Basin Pocket Mouse.

Figure 204

Total length 168 to 200 mm.; tail 88 to 107 mm.; hind foot 22 to 24 mm.; color of upper parts pale grayish buff mixed with black hairs; underparts white.

This species is the common pocket mouse in the Great Basin states. It occurs mainly in arid regions where it is abundant along roadsides, in open fields, and in sandy and rocky places. It seems to enjoy living near grain fields, much to the discouragement of the farmer.

14a. Tail longer than half total length, or at least equal to half total length 15

14b. Tail less than half total length..............................17

15a. Lower premolar larger than last molar; found mainly in California and Nevada; tail moderate in length..................16

15b. Lower premolar equal to or smaller than last molar; found mainly in Arizona; tail longer than those of 15a (more than 75 per cent of head and body length, whereas it is less than 75 per cent in *Perognathus longimembris*); body larger than that of *Perognathus longimembris*. Fig. **205**. *Perognathus amplus*. Arizona Pocket Mouse.

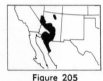

Figure 205

Total length 123 to 170 mm.; tail 72 to 95 mm.; hind foot 17 to 22 mm.; upper parts pinkish buff to pale buffy salmon, with blackish hairs throughout, so much so in some specimens (some subspecies) that they may appear blackish; underparts white or whitish.

16a. Mastoid bullae large; found only in the San Joaquin Valley of California. Fig. **206**. *Perognathus inornatus*. San Joaquin Pocket Mouse.

Figure 206

Total length 128 to 160 mm.; tail 63 to 78 mm.; hind foot 18 to 21 mm.; upper parts buff to pinkish with blackish hairs throughout; lateral line fairly well developed; underparts white; tail faintly bicolor.

16b. Mastoid bullae moderate; not found in the San Joaquin Valley. Fig. **207**. *Perognathus longimembris*. Little Pocket Mouse.

Figure 207

Total length 135 to 145 mm.; tail 75 mm.; hind foot 19 mm.; color of upper parts buff mixed with blackish hairs; underparts white.

17a. Hind foot 18 or more..18

17b. Hind foot less than 18.....................................19

18a. Color grayish olive buff. Fig. **208.** *Perognathus callistus.* **Beautiful Pocket Mouse.**

Figure 208

Total length 135 mm.; tail 63 mm.; hind foot 18 mm.

18b. Color plain buff or yellowish buff. Fig. **209.** *Perognathus apache.* **Apache Pocket Mouse.**

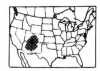

Figure 209

Total length 130 to 140 mm.; tail 64 to 68 mm.; hind foot 19 mm.

19a. Tail about 50 mm.; hind foot about 15 mm. Fig. **210.** *Perognathus flavus.* **Silky Pocket Mouse.**

Figure 210

Total length only about 115 mm.; color of upper parts pinkish buff mixed with blackish hairs; underparts white.

19b. Tail about 60 mm.; hind foot 16 to 17 mm...................20
20a. Color olivaceous. Fig. **211.** *Perognathus fasciatus.* **Olive-backed Pocket Mouse.**

Figure 211

Total length 128 to 135 mm.; tail 60 to 65 mm.; hind foot 17 mm.; color of upper parts grayish olive; underparts white.

20b. Color not olive gray.......................................21
21a. Total length about 130 mm.; tail 62 mm.; hind foot 17 mm.; color of upper parts grayish buff; underparts white. **Fig. 203.** *Perognathus flavescens.* **Plains Pocket Mouse.**

Figure 212

This pocket mouse enjoys the grassy plains of our central states where its familiar burrows may be found in roadside banks and out in the open fields.

21b. Total length 115 to 122 mm.; tail 55 to 60 mm.; hind foot 16 mm.;
color of upper parts yellowish buff mixed with black; underparts
white. Fig. 213. *Perognathus merriami.* Merriam's Pocket Mouse.

Figure 213

The Merriam Pocket Mouse lives in the low hot
valleys of the eastern part of the Southwest where
it ranges south into northern Mexico. Here it enjoys the desert plants as well as its cousin farther
north enjoys the grasses of the plains.

Genus DIPODOMYS
Kangaroo Rats

These unusual mice do not deserve the name of "rat" no matter
how one looks at them, for they are the most charming little animals
you can find. Their long tails and big hind feet, along with the small
front feet, give them the name of "kangaroo." They make friendly
and interesting pets, and are easy to capture. When you drive into
their desert habitats at night they will hop across the road in your
head light beams. All you need do is to stop and pick them up by
their convenient tails and put them into a cage. It is easy to feed
them; they will eat seeds of several kinds, especially wheat. For
moisture all they ask is a carrot, for in their desert surroundings they
probably never taste water.

Dipodomys digs extensive burrows with numerous passage ways,
storage chambers, and a nesting chamber. Here he spends his days,
coming out to forage for food at night. The prickly type of weed
seeds seem to be his preferred food. It is fortunate his cheek pockets
are fur lined.

Key to Species of Kangaroo Rats

1a. Size very large; hind foot more than 49 mm.; width of skull across
auditory bullae more than 27½ mm. 2

1b. Size medium to small; hind foot less than 49 mm.; width of skull
across auditory bullae less than 27½ mm. 4

2a. Only 4 toes present on hind feet...........................3

2b. Five toes present on hind feet. Fig. 214. *Dipodomys ingens*. Giant Kangaroo Rat.

Figure 214

Total length 310 to 350 mm.; tail 175 to 200 mm.; hind foot 50 to 55 mm.; ear about 13 to 15 mm.; color of upper parts pale dusky grayish brown, mixed with buff; dusky nose and whisker patches are not connected to form a continuous marking; cheeks whitish; rump and thigh patches washed with buff.

This big fellow can be distinguished from the other large species by his four toes; all other four-toed species are smaller in size. Found on the western side of the San Joaquin Valley, California.

3a. Dusky band across top of nose. Fig. 215. *Dipodomys spectabilis*. Banner-tailed Kangaroo Rat.

Figure 215

Total length 350 to 385 mm.; tail 210 to 285 mm.; hind foot 52 to 58 mm.; head short and broad; tail very long with a large white brush at the tip; body rather short and compact; colors light, the upper parts being yellowish buff or pinkish buff; underparts pure white.

This is the largest and most striking of the kangaroo rats, its unusually large tail and huge hind feet certainly make the name "kangaroo rat" appropriate.

3b. No dusky band across top of nose. Fig. 216. *Dipodomys deserti*. Big Desert Kangaroo Rat.

Figure 216

Total length 300 to 375 mm.; tail 180 to 215 mm.; hind foot 50 to 55 mm.; ear 12 to 15 mm.; upper parts yellowish buff; underparts pure white. The appearance of this animal is similar to that of the Banner-tailed Kangaroo Rat above, but the lack of the dark stripe over the nose and the paler colors will distinguish it.

4a. Size medium, about 240 to 340 mm. 6

4b. Size small, only about 200 to 225 mm. in total length 5

5a. Size very small, only about 200 mm. in total length; tail about 100 mm.; hind foot about 31 mm.; colors pale, almost no trace of dark markings. Found on Padre Island, Cameron County, Texas. Fig. 217. *Dipodomys compactus.* **Padre Island Kangaroo Rat.**

Figure 217

5b. Size slightly larger, about 225 mm.; tail about 150 mm.; hind foot about 35 mm.; similar to the Padre Island Kangaroo Rat. Found in southern Texas, near Brownsville. Fig. 218. *Dipodomys sennetti.* **Sennett Kangaroo Rat.**

Figure 218

6a. Width of maxillary arch at middle less than 3.9 mm. Fig. 219. *Dipodomys microps.* **Small-faced Kangaroo Rat.**

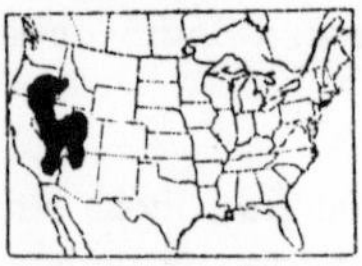

Total length 244 to 290 mm.; tail 140 to 173 mm.; hind foot 38 to 44 mm.; ear 9 to 12 mm.; colors of upper parts pale buff; dark markings much reduced; underparts white; skull very narrow or slender; desert areas of eastern California.

Figure 219

6b. Width of maxillary arch at middle more than 3.9 mm 7

7a. Total length more than 265 mm.; hind foot more than 40 mm 8

7b. Total length less than 265 mm.; hind foot less than 40 mm 16

8a. Ears more than 15 mm. in length 9

8b. Ears less than 15 mm. in length 10

9a. Color darker; color of upper parts cinnamon brown; ear rather blackish; half way toward the end the dark ventral tail stripe is

wider than the lateral white stripes. **Fig. 220.** *Dipodomys venustus.* Santa Cruz Kangaroo Rat.

Total length 306 to 332 mm.; tail 184 to 203 mm.; hind foot 44 to 47 mm.; ear 15 to 16 mm.; general colors darker than usual; five toes on hind feet.

Figure 220

9b. Color lighter; color of upper parts cinnamon buff; ear rather brownish; half way toward the end the dark ventral tail stripe is narrower than the lateral white stripes. **Fig. 221.** *Dipodomys elephantinus.* Elephant-eared Kangaroo Rat.

Total length 305 to 336 mm.; tail 183 to 210 mm.; hind foot 44 to 50 mm.; ear 16 to 18 mm.; general colors somewhat dark, but lighter than the Santa Cruz Kangaroo Rat; ears very large; toes on hind feet five.

Figure 221

10a. Four toes on hind feet. **Fig. 222.** *Dipodomys heermanni.* California Kangaroo Rat.

Total length 295 to 340 mm.; tail 182 to 217 mm.; hind foot 43 to 47 mm.; ear 14 to 15½ mm.; color of upper parts dusty buff; dark facial markings bold or pronounced; skull with small bullae.

This kangaroo rat lives in the waste lands of the western part of California where it does little if any

Figure 222

damage to crops. In fact, it eats many more grasshoppers than it does crop plants, so we can forgive this beautiful little animal if he takes a nibble here and there on our domestic plants.

10b. Five toes on hind feet......................................11
11a. Colors dark; dark facial markings pronounced; dorsal dark tail stripe at least twice as wide as lateral white stripe............12
11b. Colors medium or light; dark facial markings not pronounced; dorsal dark tail stripe less than twice as wide as lateral white

stripe. Ear 12 mm. or more; upper parts darker, warm buff or cinnamon buff. Fig. 223. *Dipodomys panamintinus.* Panamint Kangaroo Rat.

Total length 290 to 315 mm.; tail 160 to 180 mm.; hind foot 43 to 45 mm.; ear 12 to 15 mm.; toes five on hind foot; bullae moderate in size.

Figure 223

12a. **White stripe on flanks either incomplete or absent. Fig. 224. *Dipodomys morroensis.* Morro Bay Kangaroo Rat.**

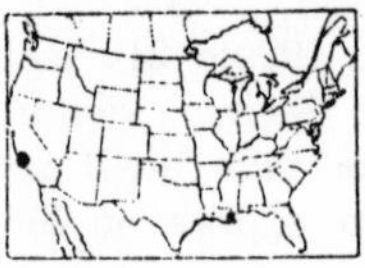

Total length 275 to 308 mm.; tail 164 to 185 mm.; hind foot 42 to 44 mm.; ear 13 to 14 mm.; color of upper parts very dark, darkest of all the kangaroo rats; no definite white band across flanks (all others have the white band much pronounced); dark facial markings are prominent. Found only at Morro Bay, California.

Figure 224

12b. **White stripe on flanks prominent**............................13

13a. **Head (and skull) long and narrow**........................14

13b. **Head (and skull) short and broad**..........................15

14a. **Four toes on hind feet; ears small; upper parts clay color. Fig. 225. *Dipodomys elator.* Loring Kangaroo Rat.**

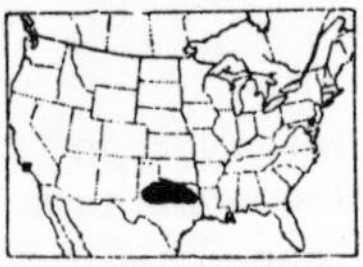

Total length about 290 mm.; tail about 170 mm.; hind foot about 44 mm.

Figure 225

14b. **Five toes on hind feet; ears large, about 14 mm.; upper parts cinnamon buff. Fig. 226. *Dipodomys agilis.* Gambel Kangaroo Rat.**

Total length 270 to 315 mm.; tail 162 to 197 mm.; hind foot 41 to 46 mm.

Figure 226

15a. Ear longer than 12.8 mm.; bullae, as observed from above, not
globular in outline. (See fig. 213.) *Dipodomys heermanni.* California Kangaroo Rat.

15b. Ear shorter than 12.8 mm.; bullae, when observed from above,
almost globular, that is, all parts of outline curved. **Fig. 227.**
Dipodomys stephensi. Stephens Kangaroo Rat.

Figure 227

Total length 277 to 300 mm.; tail 164 to 180 mm.,
hind foot 41 to 43 mm.; ear 11 to 12 mm.; color of
upper parts cinnamon buff; dark facial markings
pronounced; bullae rather large, and very much
globose in shape.

16a. Four toes on hind feet....................................17

16b. Five toes on hind feet. Fig. 228. *Dipodomys ordii.* Ord's Kangaroo
Rat.

Figure 228

Total length 220 to 245 mm.; tail 120
to 137 mm.; hind foot 35 to 40 mm.;
ear 9 to 12 mm.; color of upper parts
cinnamon buff or pale buff; dark
facial markings black and rather pronounced.

This is the kangaroo rat of the
Great Basin and Rocky Mountain
states, and it lives mainly in the
sandy sagebrush regions. It is smaller than many other species, and its
life is spent hopping about in the
sands at night in search of dry
seeds. It seems strange indeed that
these animals can exist all their lives without water, but since they
live in the damp part of the ground, and do not come out in the day
time they need far less moisture than other animals.

17a. Dark dorsal and ventral tail stripes wider than the white lateral
tail stripes. **Fig. 229.** *Dipodomys nitratoides.* San Joaquin Kangaroo Rat.

Figure 229

Total length 210 to 253 mm.; tail 120 to 152 mm.;
hind foot 33 to 37 mm.; ear 8 to 12 mm.

17b. Dark dorsal and ventral tail stripes narrower than the white lateral stripes. Fig. 230. *Dipodomys merriami*. Merriam's Kangaroo Rat.

Total length 235 to 260 mm.; tail 130 to 154 mm.; hind foot 36 to 40 mm.; ear 10 to 12 mm.; color of upper parts buff; dark facial areas quite pale.

Figure 230

Genus MICRODIPODOPS
Kangaroo Mice

These little fellows, similar in some respects to the kangaroo rats, are much smaller, lack the stripes on the tail, and always have five toes on the hind feet. They appear to be about half way between the kangaroo rats and the pocket mice, for they have greatly inflated bullae and large eyes, yet in coloration they resemble the pocket mice. They occur in the very arid regions of the Great Basin and southwestern states, where they live entirely without water, much like the kangaroo rats.

Key to the Species of Kangaroo Mice

1a. Upper parts very dark, almost blackish; top of tail tipped with black. Fig. 231. *Microdipodops megacephalus*. Dark Kangaroo Mouse.

Total length 155 to 177 mm.; tail 84 to 100 mm.; hind foot 24 to 26 mm.; weight 13 to 16 grams; color of upper parts blackish with a reddish cast; underparts almost pure white, but often with a buffy wash. The long tail and large hind feet distinguish this mouse from the pocket mice.

Figure 231

1b. Upper parts cinnamon to buffy; tail the same color throughout its length. Fig. 232. *Microdipodops pallidus.* **Pallid Kangaroo Mouse.**

Figure 232

Total length 150 to 173, mm.; tail 74 to 99 mm.; hind foot 25 to 27 mm.; weight 14 to 17 grams; color of upper parts cinnamon mixed with blackish hairs; underparts pure white.

Both species of this mouse live on seeds of desert plants, require no water, and are active throughout the year, even in very cold weather.

Family CASTORIDAE

Beavers

This family contains only one animal, the beaver, and he may be recognized at once by his large size, his webbed hind feet, and his massive, scaly, flattened tail. His teeth are very heavy, and the incisors are extremely long and sharp—so sharp that the animal is able to cut down large trees in the construction of dams across streams. Beavers work untiringly in order to construct their dams. It is almost impossible to destroy a beaver dam, for the beavers patiently repair every onslaught man may make on their construction. One must move the beavers away from the area if he does not want a dam across the stream!

Beavers live in under-water houses made of sticks and mud, or in burrows in banks of streams or ponds. They prefer freshly cut willows for food. Beavers are still one of our most valuable fur-bearing mammals. It is interesting to note (Hall, 1946, p. 482) that "Much of the early exploration of western North America by white men was incidental to their quest for beaver."

Fig. 233. *Castor canadensis.* **American Beaver.**

Figure 233

Total length about 1000 to 1200 mm.; tail about 250 to 450 mm.; hind foot about 165 to 180 mm.; color of upper parts cinnamon brown to darker brown; underparts about the same color as upper parts.

Beavers are the largest rodents in our country. They have been exterminated in some parts of the United States, but are very common in others, and are constantly being introduced by state game departments into new territories. They do considerable damage to trees along streams, and to ditch banks. There are even records of beaver cutting down trees on the front lawns of city dwellers. But on the whole the beaver is to be encouraged in our country, for it is a valuable asset to the maintenance of natural conditions in the mountains, it is a valuable flood control worker, and a builder of ponds which harbor all sorts of other animals and plants which help to make our country interesting.

Family CRICETIDAE
New World Mice and Rats

These interesting little animals are shunned by many people because of long association with the common European mice and rats which frequent our houses. Actually, the American mice and rats form a separate family. They seldom enter houses, and cause little damage even when they do—with the exception of the wood rat or pack rat. These animals lack the characteristic odor of the "domestic" mice and rats, and seldom carry the diseases that the European animals carry.

Key to the Genera of Native Mice and Rats

1a. Upper molars with the biting surface a series of tubercles rather than low flat ridges...2

1b. Upper molars with the biting surface a series of low flat ridges...6

2a. Tubercles on upper molars in the two rows; fur usually soft......3

2b. Tubercles on upper molars in flattened S-shaped loops; fur rough to the touch. *Sigmodon*, page 140.

3a. Tail more than half the length of head and body...............4

3b. Tail less than half the length of head and body. *Onychomys*, page 135.

4a. Upper incisors with longitudinal grooves. *Reithrodontomys*, page 129.

4b. Upper incisors not grooved....................................5

5a. Tail naked and scaled; fur coarse; belly not pure white. *Oryzomys*, page 140.

5b. Tail haired, not scaled; fur soft; belly pure white. *Peromyscus*, page 131.

6a. Tail more than half as long as head and body; body large and slender. *Neotoma*, page 136.

6b. Tail usually much less than half as long as head and body; body small 7

7a. Size large, 300 to 600 mm.; tail long, 110 to 275 mm.; hind foot 40 to 90 mm. 8

7b. Size small, much less than 300 mm. 9

8a. Tail round, size over 300 mm., but much smaller than the common muskrat. *Neofiber*, page 144.

8b. Tail flattened laterally, size over 500 mm., much larger than the Round-tailed Muskrat. *Ondatra*, page 143.

9a. Incisors with shallow groove in outer surface. *Synaptomys*, page 141.

9b. No groove in incisors......................................10

10a. Angles of enamel ridges of lower molars much deeper on outer side than on inner side; angles of upper molars about equal. *Phenacomys*, page 144. (See fig. 265.)

10b. Angles about equal on molars of both jaws..................11

11a. Fur of back reddish; molars with roots or prong-like projections. *Clethrionomys*, page 142.

11b. Fur not reddish; molars without roots......................12

12a. Tail very short, seldom much more than length of hind foot; ears partly concealed in fur....................................13

12b. Tail usually short, but always longer than hind foot; ears plainly visible. *Microtus*, page 146.

13a. Mammae 4; fur dense and short, almost mole-like; found in the eastern states. *Microtus* (in part), pages 149 (10b) and 150 (11a, 11b).

13b. Mammae 8; fur long and lax, not mole-like; found in the western states. *Lagurus*, page 152.

Genus REITHRODONTOMYS

Harvest Mice

These little fellows look somewhat like the common white-footed mouse *(Peromyscus)*, or even like a house mouse *(Mus)*, but they have a buff line along the side of the body, which is almost orange in some specimens. Their feet are always white, which will distinguish them from the house mice, while their dusky or buffy belly will distinguish them from the white-footed mice. They are always smaller than the white-footed mice.

Key to the Species of Harvest Mice

1a. Size small; total length about 120 mm.; tail about 55 to 60 mm.; hind foot 16 mm.; color of upper parts dark brown. Fig. 234. *Reithrodontomys humulis.* Eastern Harvest Mouse.

Figure 234

Harvest mice live in the waste patches of weeds and litter along the roadsides, in unkept briar patches, or in other places where weeds and brush become hopelessly tangled. They feed on weed seeds and grain, and do some damage to grain crops.

The distribution of this species in the southern states is in five distinct groups, each one being a different subspecies. Their absence throughout most of Georgia and eastern Alabama seems strange, but further collecting in that area may prove that they occur there, too.

1b. Size larger; color paler than dark brown........................2

2a. Total length about 175 mm.; tail about 100 mm.; hind foot about 20 mm.; color of upper parts yellowish buff. Fig. 235. *Reithrodontomys fulvescens.* Harvest Mouse.

Figure 235

This little harvest mouse lives in the western part of the southern Mississippi River valley and west into Texas, then south into Mexico, where it is able to live in extremely arid regions, although it prefers river valleys to arid deserts.

2b. Size smaller 3

3a. Upper part of body without middorsal stripe; total length about 135 to 150 mm.; tail about 65 to 75 mm.; hind foot about 17 to 18 mm.; color of upper parts brownish buff. Fig. 236. *Reithrodontomys megalotis.* Western Harvest Mouse.

Figure 236

Upper parts brownish buff or sometimes paler; underparts usually washed with buff, and the ground color usually gray. Habits similar to Eastern Harvest Mouse.

A closely related species, *Reithrodontomys raviventris*, occurs in the salt marshes around San Francisco Bay, California.

3b. Upper part of body with dark middorsal stripe; total length about 107 to 143 mm.; tail about 48 to 63 mm.; hind foot 14 to 17 mm.;

color pale gray to buffy gray washed with fulvous, and with underparts white. *Reithrodontomys montanus.* **Plains Harvest Mouse.**

This mouse is difficult to distinguish from 3a above where the two are found in the same locality, but the lack of the middorsal stripe in 3a is the best way to tell the two apart. The shorter tail of this mouse is also a helpful point to note, as is the smaller foot. This mouse lives from Eastern Wyoming and South Dakota south to Texas and west to southern Arizona, including all of New Mexico and eastern Colorado.

Genus PEROMYSCUS

White-footed or Deer Mice

White-footed mice are the most abundant mammals in North America, and may be found in almost every habitat, from under piled driftwood along the sea coast to the tops of the highest mountains. There is, in fact, an authentic record of one *Peromyscus* being found at the top of Mount Rainier in the state of Washington, nearly 9000 feet above the nearest vegetation, and with nothing but ice and snow on all sides.

It is easy to recognize this mouse, for the tail is usually rather long, the feet are white or pink, the belly is almost always pure white, and the upper parts are dusky gray in young specimens, and buffy brown in adults, with a browner area along the sides of the body. The eyes are large and beady, and the ears fairly long.

Key to the Species of White-footed Mice

1a. Five plantar tubercles on hind feet; total length 186 to 221 mm.; tail 80 to 95 mm.; hind foot 24 to 29 mm.; ear 22 to 25 mm. *Peromyscus floridanus.* Florida Mouse.

This rather rare mouse is found only in Florida, and seems to live in sandy places where the soil is well drained, especially at the tops of ridges, in palmettos and among pines.

1b. Six plantar tubercles on hind feet; sizes vary.................2
2a. Total length more than 185 mm...............................3
2b. Total length less than 185 mm...............................10
3a. Ear shorter than hind foot..................................4
3b. Ear longer than hind foot...................................5
4a. Hind foot 24 mm. or more...................................6
4b. Hind foot less than 24 mm..................................7

5a. Ear about 24 mm. in length from notch. Fig. 237. *Peromyscus truei.* **Piñon Mouse.**

Total length 177 to 212 mm.; tail 87 to 98 mm.; hind foot 22 to 27 mm.; ear 24 to 27 mm.; weight 20 to 29 grams; color of upper parts buff with some dusky mixed in; underparts white. These mice prefer rocky areas among pinon pines and junipers.

Figure 237

5b. Ear about 20 mm. in length from notch.   Fig. 238. *Peromyscus nasutus.* **Long-nosed Deer Mouse.**

Total length about 188 mm.; tail about 92 mm.; hind foot about 24 mm.; upper parts grayish brown; underparts white; mammae 6. .This mouse occurs in rocky areas at moderate elevations in the southwest.

Figure 238

6a. Size very large, total length 225 mm. or more; tail 125 mm. or more. Fig. 239. *Peromyscus californicus.*

Hind foot about 27 mm.; color of upper parts dusky reddish brown; underparts white.   This is the largest of all the white-footed mice, and may reach the length of 250 mm. with a tail of nearly 140 mm.; mammae 4.   This mouse inhabits brushy areas along the California coast.

Figure 239

6b. Size about 190 to 200 mm.; tail 85 to 90 mm. Fig. 240. *Peromyscus floridanus.* **Florida Deer Mouse, or Gopher Mouse.**

Hind foot about 26 mm.; color of upper parts yellowish buff to light brownish buff; underparts white. This big-footed mouse inhabits the dry ridges where black-jack oak is common.

Figure 240

7a. Hind foot about 23 mm.; found in Texas. Fig. 241. *Peromyscus pectoralis*.

Figure 241

Total length about 187 mm.; tail about 96 mm.; upper parts buff mixed with dusky; underparts white. A resident of the hot, semi-deserts of west-central Texas and south into Mexico, where it lives among the cactus and mesquite.

7b. Hind foot less than 23 mm., or if 23 mm., then not found in Texas. . 8

8a. Tail with a tuft of hair at the tip; tail usually longer than half of the total length. 9

8b. Tail without a tuft of hair at the tip; tail usually shorter than half of the total length. Fig. 242. *Peromyscus maniculatus*.

Figure 242

Total length 160 to 220 mm.; tail 65 to 120 mm.; hind foot 19 to 21 mm.; mammae 6; upper parts buffy brown to dusky brown, or even darker; underparts white.

This is the most abundant species of *Peromyscus*, and has a very wide range as the map will show. It occurs almost everywhere from the hot desert valleys to the highest mountains, often occuring in very large numbers. Specimens of this little mouse have been found nearly to the summit of Mt. Rainier in the state of Washington, 9000 feet above the zone of vegetation. There are about 30 subspecies in all.

9a. Sole of hind foot naked; mammae 4; ear 18 to 20 mm.; upper parts light in color; weight about 24 grams. Fig. 243. *Peromyscus eremicus*.

Figure 243

Total length 187 to 214 mm.; tail 92 to 113 mm.; hind foot 20 to 22 mm.; color of upper parts buff overlaid with a dusky wash; underparts white. This mouse occurs in the low hot valleys of the Southwest, never extending into the timbered areas.

9b. Nearly first half of sole of hind foot hairy; mammae 6; ear 19 to 21 mm.; upper parts very dark in color; weight about 32 grams. Fig. 244. *Peromyscus boylii*.

Total length 197 to 211 mm.; tail 98 to 113 mm.; hind foot 22 to 24 mm.; upper parts dark brown to blackish, buffy along the sides; underparts whitish; some specimens are also buffy on the upper parts. This species occurs in brushy areas of the mountains of the Southwest.

Figure 244

10a. Size less than 140 mm.....................................11

10b. Size more than 140 mm.....................................12

11a. Size about 100 mm.; tail about 40 mm.; hind foot 14 mm. Fig. 245. *Peromyscus taylori.*

Color of upper parts pale grayish; underparts smoke gray, not white. This is the smallest of the white-footed mice.

Figure 245

11b. Size about 120 to 140 mm.; tail 40 to 50 mm.; hind foot 16 to 17 mm. Fig. 246. *Peromyscus polionotus.* Beach Mouse.

Color of upper parts brownish buff, paler than most of the white-footed mice; underparts creamy white. It inhabits the sandy fields and beaches of the southern states, seldom being found in wooded regions.

Figure 246

12a. Found from the Rocky Mountains to Oregon and California, and south to Mexico; mammae 4; fur very long and lax; color of upper parts yellowish buff mixed with gray. Fig. 247. *Peromyscus crinitis.* Canyon Mouse.

Total length 162 to 186 mm.; tail 79 to 101 mm.; hind foot 18 to 21 mm.; ear 17 to 21 mm.; color of upper parts pale buffy gray; underparts whitish.

This mouse lives among rocks in canyons, especially where mountain canyons open out into valleys.

Figure 247

12b. Found in central and eastern states; mammae 6; fur shorter; color not pale buffy gray..13

13a. Upper parts rich tawny, almost golden; hind foot 18 to 20 mm.; underparts creamy white, often yellowish. **Fig. 248.** *Peromyscus nuttallii.* **Golden Mouse.**

Figure 248

Total length 158 to 186 mm.; tail 68 to 88 mm. This mouse occurs in a variety of habitats in the southern states, in the mountains as well as in the valleys. It is a good climber, and may be often found high in trees. It may make its nest in crotches of bushes or trees.

13b. Upper parts not tawny, hind foot 19 to 22½ mm.; underparts white or dull white...14

14a. Upper parts grayish brown to reddish brown; underparts pure white; hind foot 20 mm. **Fig. 249.** *Peromyscus leucopus.* **Wood Mouse.**

Figure 249

Total length 152 to 181 mm.; tail 59 to 83 mm.; hind foot 19½ to 22 mm. Commonly found in the deep woods of the eastern states, seldom occurring in the open. Occasionally it comes into houses, making stores of nuts in various places about the house.

14b. Upper parts very dark, dusky brown; underparts dull white; hind foot about 22 mm. **Fig. 250.** *Peromyscus gossypinus.* **Cotton Mouse.**

Figure 250

Total length 164 to 192 mm.; tail 71 to 87 mm.; hind foot 21 to 22½ mm.; similar to *P. leucopus*, but larger and much darker in color. This mouse occurs in a wide variety of habitats in the southern states; along the marshy coastal areas of the south Atlantic states; in the foothills at moderate elevations; in low swampy areas along the Mississippi. The common name implies that it lives in cotton fields, but strangely enough, such a habitat seems to be the exception rather than the rule.

Genus ONYCHOMYS
Grasshopper Mouse

This mouse looks very much like a white-footed mouse, but has an unusually short tail. It lives only in the arid regions of the west where it is a wanderer among the sagebrush plains of the Great Basin states. It seldom remains for long in any one place, except when rearing the young. This mouse eats some plant material, but insects and other small animals make up such a large percent of its food that it has rightly earned its common name. Scorpions, crickets, beetles, small lizards, and even other mice (especially those caught in a mammalogist's traps) form part of the diet of this interesting mouse.

Key to the Species of Grasshopper Mice

1a. Tail usually less than half as long as head and body; size somewhat larger; third molar as long as broad; first molar less than half the length of the tooth row. Fig. 251. *Onychomys leucogaster.* Northern Grasshopper Mouse.

Total length 131 to 151 mm.; tail 35 to 43 mm.; hind foot 19 to 21½ mm.; ear about 18 mm.; color of upper parts grayish buff; sides somewhat reddish; underparts white; feet and underside of tail white.

These mice are difficult to trap intentionally, for they never remain in one place very long. If one is caught one night in a certain spot, it is not likely that another will be caught near there again.

Figure 251

1b. Tail usually more than half as long as head and body; size somewhat smaller; third molar broader than long; first molar more than half the length of tooth row. Fig. 252. *Onychomys torridus.* Southern Grasshopper Mouse.

Total length 126 to 150 mm.; tail 35 to 52 mm.; hind foot 18 to 20 mm.; weight 20 to 25 grams; color of upper parts light pinkish cinnamon, somewhat washed with dusky; underparts pure white. This mouse wanders through the low hot valleys farther south where creosote bush, cactus and mesquite are common plants.

Figure 252

Genus NEOTOMA

Wood Rats

These interesting fellows are the curious, pestherous rats which one encounters when he goes camping. Whether you sleep in your sleeping bag out in the desert, or in a mountain cabin, you will sooner or later make acquaintance with the wood rat. He is the creature who wakes you up in the middle of the night as he packs off your silverware or your fountain pen. If you turn the flashlight on him he will likely just stare at you or stamp his hind foot in disapproval. But if you can overlook his idiosyncrasies you will find him quite fascinating. He usually has a rather long tail, either bushy or only scantily haired, very long wiskers, large beady eyes, and a fairly heavy body. The size varies greatly from the large species of the Pacific Northwest to the small desert wood rats of the Southwest.

Key to the Species of Wood Rats

1a. Tail large and bushy, almost squirrel-like, and slightly flattened; hairs on tail 20 mm or more in length. **Fig. 253. Neotoma cinerea.** Bushy-tailed Wood Rat.

Total length 350 to 435 mm.; tail 140 to 207 mm.; hind foot 40 to 52 mm.; ear 30 to 44 mm.; color of upper parts grayish to blackish brown; underparts white or dingy white, with a dirty yellow spot in the center of the belly; feet white or whitish.

Wood rats occur in both wooded regions and in rocky arid regions over most of the western states, except the extreme Southwest. Their nests are located in rocky crevices, under logs, in hollow stumps, or in cabins.

Figure 253

1b. Tail not bushy, round, tapering to a sharp point; hairs much less than 20 mm. in length...2

2a. Total length less than 320 mm. **Fig. 254. Neotoma lepida.** Desert Wood Rat.

Total length 256 to 312 mm.; tail 110 to 128 mm.; hind foot 29 to 34 mm.; weight 120 to 188 grams; color of upper parts buff to blackish buff; underparts whitish to buffy. This is the wood rat of the low arid regions of the Southwest, occurring mainly in rocky hills, but occasionally inhabiting rather high elevations.

Figure 254

A closely related species, *Neotoma stephensi,* occurs in Utah, Arizona and New Mexico. The tail is somewhat more hairy than in *Neotoma lepida.*

2b. Total length 320 mm. or more...................................3

3a. Total length 320 to 360 mm...................................4

3b. Total length 365 to 450 mm·...................................7

4a. Hind foot 30 to 32 mm.; upper parts grayish brown; underparts

white; feet white; tail black above, dull white beneath. Fig. 255. *Neotoma intermedia.* Wood Rat. Total length about 325 mm.; tail about 160 mm.

Figure 255

4b. Hind foot 33 to 41 mm..5

5a. Hind foot 33 to 37 mm.......................................6

5b. Hind foot 38 to 41 mm.; upper parts pale drab; underparts white;

tail blackish above, but gray beneath· Fig. 256. *Neotoma micropus.*

Total length about 350 mm.; tail about 165 mm.

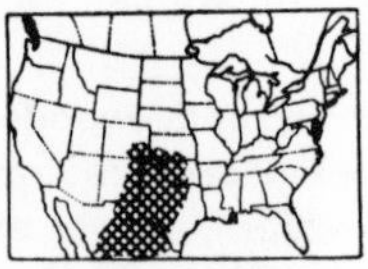

Figure 256

6a. Upper parts pinkish buff; underparts pure white; upper part of tail brownish.   Fig. 257.   *Neotoma albigula.*

Total length about 325 mm.; tail about 150 mm.

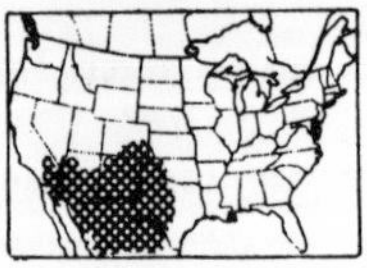

Figure 257

6b. Upper parts grayish buff; underparts dull white; upper part of tail black.   Fig. 258.   *Neotoma mexicana.*

Total length about 325 mm.; tail about 150 mm.

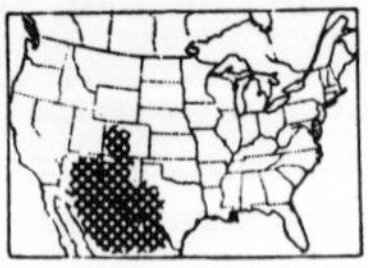

Figure 258

7a. Tail more than 200 mm.; total length 425 to 450 mm.; upper parts buff with a dusky area in the center of back; underparts cream to pinkish buff; throat and chest pure white; forefeet and toes of

hind feet white; rest of hind feet dusky. **Fig. 259.** *Neotoma fuscipes.* **Dusky-footed Wood Rat.**

These wood rats are famous for their enormous structures of sticks and rubbish which they make for nests. They prefer their own massive piles of sticks to rocky crevices or cabins.

Figure 259

7b. Tail 200 mm. or less; found in the eastern states.................8

8a. Hind foot 40 to 46 mm.; upper parts grayish buff; tail dark above and white below. **Fig. 261.** *Neotoma magister.* **Cave Rat.**

Total length 405 to 440 mm.; tail 170 to 200 mm.; found from Tennessee to Connecticut, where it inhabits the rock slides in the mountainous areas, where there is very little vegetation.

Figure 260

8b. Hind foot 36 to 40 mm.; upper parts cinnamon; tail the same color above and below. **Fig. 261.** *Neotoma floridana.* **Florida Wood Rat.**

Total length 362 to 409 mm.; tail 166 to 189 mm.; found in the southern states, where it inhabits a variety of regions, from swamps to rock piles. It often builds nests of sticks and shredded plants, placed frequently in the branches of osage orange trees. These wood rats eat berries and juicy plants, and even hickory nuts and pawpaw seeds.

Figure 261

Genus ORYZOMYS
Rice Rats

Fig. 262. *Oryzomys palustris.* **Rice Rat or Marsh Rat.**

This little rat is long and slender in form, with a long tail and coarse fur; upper parts dark brown; underparts light brown to buff, or grayish white; never pure white.

Total length about 225 to 275 mm.; tail about 115 to 150 mm.; hind foot about 30 mm. to 35 mm.

These little fellows live in marshy areas along the sea coast or even in clearings and brushy areas in the mountains, but ordinarily prefer damp marshy places where they make runways through the marsh grass. They are good swimmers, and may dive under the water if threatened by enemies.

Figure 262

Genus SIGMODON
Cotton Rats

These rats look somewhat like the rice rats, but are larger and with even coarser fur. They are smaller than most species of wood rats, and their fur is more coarse. The ears are large, tail long, molars with crowns flattened and showing S-shaped loops.

Key to Species of Cotton Rats

1a. **Size small, total length about 185 mm.; tail about 95 mm.; hind foot about 28 to 31 mm. Fig. 263.** *Sigmodon minimus.* **Least Cotton Rat.**

Color of upper parts grizzled gray; under parts buff.

These rats make runways through the grass in damp meadows, somewhat similar to those of meadow mice *(Microtus).* They may also live in drier habitats where their trails run from one clump of grass to another.

Figure 263

A closely related species, *Sigmodon ochrognathus,* occurs in southern Arizona, southern New Mexico and southern western Texas. It is somewhat larger than *Sigmodon minimus.*

1b. Size larger, total length 215 to 285 mm.; tail 75 to 110 mm.; hind foot 26 to 33 mm. Fig. 264. *Sigmodon hispidus.* Cotton Rat.

Figure 264

Color of upper parts grizzled buff mixed with many blackish hairs; underparts pale gray to buff; tail scaly, very sparsely haired; fur long and coarse.

This rat lives in open fields and in wet meadows, up to as high as 1700 feet elevation. They make trails or extensive runways among the grass and weeds, where they are active both day and night.

Genus SYNAPTOMYS
Lemming Mice

These small, meadow-mouse like rodents live mainly in the very damp meadows of the mountains of the northeastern states and Canada. One species is confined to high mountain meadows in the United States, while the other occurs both in meadows and on dry hillsides, even down into wooded valley areas. The extremely short tail and the grooved incisors should distinguish this mouse from all others.

Key to Species of Lemming Mice

1a. Lower molars with triangles on outer side (fig. 265); found in the central and eastern states. Fig. 266. *Synaptomys cooperi.* Lemming Mouse or Bog Lemming.

OUTER   SIDE
Figure 265

Figure 266

Total length 114 to 130 mm.; tail 13 to 18 mm.; hind foot 17 to 18 mm.; weight 24 to 35 grams; color of upper parts dark brownish gray, somewhat grizzled in appearance; underparts silvery gray. These little mice live in a variety of habitats, generally in company with other small mammals, inhabiting the runways and tunnels of other mice, and even shrews.

1b. **Lower molars without triangles on outer side** (fig. 267); found in the state of Washington, and northward into Canada. Fig. 268. *Synaptomys borealis*. Lemming Mouse.

OUTER SIDE

Figure 267

Figure 268

Total length about 120 mm.; tail about 25 mm.; hind foot 18 mm.; color of upper parts varying from dark brown to dark gray; underparts somewhat lighter. These mice live in high damp mountain meadows on the slopes of Mount Baker, Washington, and probably on other high peaks in the northern ranges.

Genus CLETHRIONOMYS
Red-backed Mice

These mice look very much like meadow mice *(Microtus)*, but the fur on the back is reddish in color. The molars have root-like prongs on them, and have no cement in their reentrant angles. They live in forested areas of the northern states and Canada. They feed on green plants, roots, and lichens, storing away quantities of such provisions for winter use.

Key to Species of Red-backed Mice

1a. **Color very dark,** often so dark that the reddish color of the back is obscured; tail longer, about 50 mm. Fig. 269. *Clethrionomys occidentalis*. Western Red-backed Mouse.

Figure 269

Total length about 160 mm.; tail about 50 mm.; hind foot about 21 mm.; color of upper parts dark brown with the reddish color almost chestnut. These mice are found about fallen logs or stumps near streams in the coniferous forests of the coastal areas.

1b. Color varying, but never concealing the bright reddish color of the back; tail shorter, less than 50 mm. **Fig. 270** *Clethrionomys gapperi.* **Red-backed Mouse.**

Total length 140 to 156 mm.; tail 35 to 48 mm.; hind foot 18 to 19 mm.; color of upper parts rufous in the middle of the back, with the surrounding areas grayish to grayish brown; underparts grayish white. This species inhabits very damp, mossy areas in mountain forests, where they often make short runways under moss and lichens.

Figure 270

Genus ONDATRA

Muskrats

Fig. 271 *Ondatra zibethicus.* **Common Muskrat.**

The common muskrats are large rodents with their hind feet partly webbed. Their tail is nearly as long as the body, flattened laterally, with only a few hairs. Eyes and ears are small. The fur is very thick and oily, "waterproofed" for the muskrat's aquatic existence.

Total length 500 to 600 mm.; tail 200 to 300 mm.; hind foot 80 to 86 mm.; ear 20 to 24 mm.; color of upper parts dark brown to blackish; underparts buff to cinnamon brown.

Muskrats occur in nearly all the waterways of America, whether it be small streams or ponds. In streams they make their nests in holes in banks, while in many lakes they construct mounds of floating rushes which serve as a nest in both winter and summer. As a rule the muskrat is a vegetarian, though occasionally one is known to eat water animals. Succulent tule stems and rushes are preferred food.

Figure 271

Genus NEOFIBER

Round-tailed Muskrat

Fig. 272. *Neofiber alleni.* **Round-tailed Muskrat, or Water Rat.**

Figure 272

This muskrat is much smaller than the common muskrat; its tail is round, and much less than half of the total length.

Total length 300 to 330 mm.; tail 115 to 130 mm.; hind foot about 44 mm.; color of upper parts dark brown; underparts buff to almost white; fur thick and oily.

The habits of this muskrat are like those of the common muskrat. It occurs from southern Georgia to Florida.

Genus PHENACOMYS

Phenacomys

These animals look very much like ordinary meadow mice, but the inner reentrant angles of the enamel of the lower molars are much deeper than the outer angles (fig. 273). The molars are also rooted, rather than rootless as are those of the meadow mice. The fur is more lax, softer and longer. They are found almost entirely in mountains, where they often nest in heather high up near the snow fields.

INNER ANGLES

Figure 273

Key to the Species of Phenacomys

1a. Tail short, only 30 to 45 mm.; color of upper parts buffy gray to dusky gray. Fig. 274. *Phenacomys intermedius.*

Total length 145 to 181 mm.; tail 34 to 55 mm.; hind foot 16 to 20 mm.; color of underparts whitish. These mice live along small mountain streams near timberline where they can jump across from stone to stone. Here they may make very short runways among the moss and alpine plants.

Figure 274

1b. Tail longer, about 60 to 85 mm.; color brown or reddish.........2

2a. Color of upper parts reddish; tree-dwelling in habits. Fig. 275. *Phenacomys longicaudus.* Red Tree Mouse.
Total length about 180 mm.; tail about 75 mm.; hind foot 21 mm.; underparts buffy white.

These are among the most peculiar mice in the world, for they spend nearly all their time in the tops of the large Douglas Fir trees. They make large ball-like nests of fir needles and shredded bark, and almost never come to the ground. Their food consists largely of fir needles.

Figure 275

2b. Color brown or grizzled brown...............................3

3a. Feet gray; color brown; tree-dwelling in habits. Fig. 276. Phena-comys silvicola. Dusky Tree Mouse.

Figure 276

Total length about 190 mm.; tail about 80 mm.; hind foot about 22 mm.; upper parts dark brown; underparts brown washed with whitish.

This tree mouse is even more rare than the Red Tree Mouse. Its habits are similar, except that it lives mainly in Sitka spruce and hemlock trees rather than in Douglas firs.

3b. Feet white; color brown to grizzled brown; not found in trees. Fig. 277 Phenacomys albipes. White-footed Phenacomys.

Total length about 165 to 170 mm.; tail about 60 mm.; hind foot about 19 mm.; upper parts brown; underparts buffy.

Figure 277

This is another very rare mouse, which inhabits the deep forests along the coast of Oregon and Northern California, where it may be found on fallen logs and about stumps. Little is known about its habits.

Genus MICROTUS
Meadow Mice

These are the very common meadow mice that live in nearly all the meadows and grassy areas of America, from the high mountains to the lowest coastal plains. They remain active above ground all year, making extensive runways under cover of grass and other plants. They make subsurface burrows as well, and it is down in these that they place their nests of dried grass. They feed on green plants, possibly supplemented by bark in the winter.

Their molars do not have prong-like roots, and the incisors are not grooved. Unlike *Lemmiscus*, the tail is never shorter than the hind foot.

Key to the Species of Meadow Mice

1a. Plantar tubercules 5. Fig. 278 2

Figure 278

1b. Plantar tubercles 6. Fig. 279. 7

Figure 279

2a. Size medium to large, total length 125 to 260 mm.; if found in South Dakota, North Dakota, Minnesota, and adjacent Canada, then the total length is more than 130 mm.; and the tail more than 35 mm. 3

2b. Size very small, total length 120 to 130 mm.; found only in South Dakota, North Dakota, Minnesota, and adjacent Canada. Fig. 280. *Microtus minor.*

Total length 120 to 133 mm.; tail 30 to 36 mm.; hind foot 16½ mm.; color of upper parts grizzled gray; underparts lighter, washed with buff.

Figure 280

3a. Size large, total length 180 to 260 mm.; tail 60 to 90 mm.; hind foot 26 to 30 mm.; mammae 8. Fig. 281. *Microtus richardsoni.* Big-footed Water Vole.

Color of upper parts grayish brown; underparts pale gray to whitish.

This large meadow mouse may be found in the high mountain meadows in the Rockies and the Cascades. Here it builds its giant runways among the heather along the small meadow streams flowing out from under snowbanks. Runways often cross the streams and go straight up the bank on the other side only to disappear again in the dense meadow cover.

Figure 281

3b. Size smaller, total length 125 to 180 mm.; tail 25 to 47 mm.; hind foot 16 to 22 mm.; mammae 6 or 8 4

4a. First lower molar with 3 closed and 2 open triangles; third upper molar with 2 closed triangles (fig. 282); mammae 6; not found in the Pacific Northwest...............5

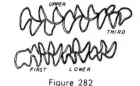

Figure 282

4b. First lower molar with 5 closed triangles; third upper molar with 3 closed triangles (fig. 283); mammae 8; found in the Pacific Northwest. Fig. 284. *Microtus oregoni.* Oregon Meadow Mouse.

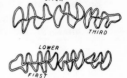

Figure 283

Figure 284

Total length 125 mm. to 161 mm.; tail 25 to 46 mm.; hind foot 16 to 22 mm.; color of upper parts blackish brown; underparts dark gray with a buffy wash. These little mice are the common meadow mice in the grassy lowlands, the foothills, and even in some of the high mountain meadows of the coastal area.

5a. Found in dry prairies in southern Montana, Wyoming, eastern Colorado, South Dakota, Nebraska, and Kansas. Fig. 285. *Microtus haydeni.* Hayden Vole.

Figure 285

Total length about 180 mm.; tail about 47 mm.; hind foot about 22 mm.; color of upper parts light gray; underparts silvery white. These mice are found in very dry situations, among sagebrush on dry hillsides, or in badlands.

5b. Found east of the range of *Microtus haydeni*....................6

6a. Found in southern Louisiana. Fig. 286. *Microtus ludovicianus.* Louisiana Meadow Mouse.

Figure 286

Total length 133 to 145 mm.; tail 28 to 38 mm.; hind foot 17 to 19 mm.; color of upper parts dark gray, nearly black, mixed with dark brown; underparts dark buff.

6b. Found from eastern Kansas and eastern Nebraska to Indiana. Fig. 287. *Microtus ochrogaster.* Prairie Meadow Mouse.

Figure 287

Total length 134 to 162 mm.; tail 25 to 37 mm.; hind foot 17 to 22 mm.; color of upper parts grizzled gray; underparts buff. This mouse is common in the prairie states where it lives mainly in moist places; it may sometimes be found in dry situations as well.

7a. Second upper molar with 4 closed sections and a rounded posterior loop (fig. 288); color dark brownish gray, even blackish; size usually rather large 8

Figure 288

7b. Second upper molar with 4 closed sections, but no posterior loop (except sometimes in *Microtus californicus*) (fig. 289); color either dark or light; size varying.................9

Figure 289

8a. Third upper molar with 2 of the 3 triangles confluent (see Fig. 290); colors paler than in *Microtus pennsylvanicus*, about buffy gray; found only on Muskeget Island, Massachusetts. *Microtus breweri*. Beach Vole.

Figure 290

Total length 153 to 191 mm.; tail 45 to 60 mm.; hind foot 22 to 23 mm.; color of upper parts buffy gray; underparts silvery gray.

8b. Third upper molar with 3 closed triangles. Figs. 291 and 292 *Microtus pennsylvanicus*. Eastern Meadow Mouse.

Figure 291

Figure 292

Total length 145 to 185 mm.; tail 36 to 55 mm.; hind foot 19 to 21 mm.; color of upper parts chestnut brown to very dark blackish brown; underparts grayish often washed with pale cinnamon. This very common mouse lives in damp meadows where it builds its maze of runways among the grass.

9a. Mammae 4 in number.......................................10
9b. Mammae 8 in number.......................................12

10a. Size small, about 90 to 115 mm. total length.................11

10b. Size larger; colors dull brown or plain gray above. *Microtus mexicanus*. Mexican Vole.

Total length, 121 to 152 mm.; tail 24 to 35 mm.; hind foot, 17 to 21 mm.; ear, 12 to 15 mm. This Mexican species extends into the Southwestern states as well, and southward to southern Mexico in the mountains.

11a. Size smallest, total length less than 100 mm.; tail only 15 mm.; hind foot 14 mm.; color tawny above; underparts dusky. Fig. 293 a *Microtus parvulus*. Florida Pine Vole.

Figure 293 a

Total length about 95 mm.; color of upper parts light brown or tawny; underparts drab. These little fellows make extensive runways beneath the soil under pine trees, where they live much like moles. They may even use runways of moles to some extent. They come to the surface more often than do moles.

11b. Total length 110 to 115 mm.; tail 17 to 19 mm.; hind foot 15 to 16 mm.; underparts dusky. Fig. 293 b *Microtus pinetorum*. Pine Vole.

Figure 293 b

Color of upper parts brown to chestnut; underparts dusky to silvery gray. This mouse does not live in pine woods, but prefers the open fields or wooded areas where it lives just beneath the surface, making runways under the dead leaves and other litter on the ground. Most of the time is spent under ground, for it comes to the surface only to run to another burrow a few inches away.

12a. A pair of glands on the flanks of males (can be seen after skin is removed); nose yellowish. *Microtus chrotorrhinus*. Rock Vole.

Total length, 137 to 170 mm.; tail, 45 to 50 mm.; hind foot, 19 to 22 mm. Found only in eastern Canada north of the Great Lakes (including N E Wisconsin), in the mountains of New England, and south in the mountains to Tennessee.

12b. A pair of glands on the hips of males, not on the flanks (may be seen after skinning); nose not yellowish.................13

13a. Tail usually longer than 60 mm.............................14

13b. Tail less than 60 mm.......................................15

14a. Hind foot usually 19 to 24 mm.; total length usually 160 to 200 mm.; color usually gray. **Fig. 294.** *Microtus longicaudus.* **Long-tailed Meadow Mouse.**

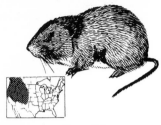

Total length 160 to 230 mm.; tail 55 to 100 mm.; color of upper parts brownish gray; underparts dull whitish. Unlike most *Microtus,* these mice seldom seem to build runways, preferring to forage along small streams, or to use runways of other meadow mice.

Figure 294

14b. Hind foot usually 26 to 30 mm.; total length usually 180 to 240 mm.; color usually blackish; one subspecies found in the San Juan Islands of Washington is smaller, with the tail much shorter. **Fig. 295.** *Microtus townsendii.* **Townsend Meadow Mouse.**

Total length 160 to 245 mm., but usually over 200 mm.; tail 50 to 96 mm., but usually over 65 mm.; hind foot usually 26 to 30, but may be less in some subspecies; color of upper parts blackish brown; underparts grayish. The runways are very large in diameter.

Figure 295

15a. Found in the valleys and along the coast of California, and into western Oregon; incisive foramina (fig. 296) wider at posterior end; incisors abruptly curved. **Fig. 297.** *Microtus californicus.* **California Meadow Mouse.**

INCISIVE FORAMINA

Figure 296

Total length about 170 mm.; tail about 50 mm.; hind foot about 21 mm.; color of upper parts dull buffy brown; underparts buffy gray or soiled whitish.

Figure 297

15b. Found in the western states east of the Cascade Mountains (Oregon and Washington), and in states east of California; does not occur in the range of *Microtus californicus*; incisive foramina (fig. 298) narrower at posterior end; incisors not abruptly curved.   Fig. 299.   *Microtus montanus*.   Gray Meadow Mouse.

Figure 298

Figure 299

Total length 130 to 175 mm.; tail 28 to 50 mm.; hind foot 17 to 21 mm.; color of upper parts grizzled gray to grizzled brownish gray; underparts whitish.

This mouse occurs in meadows as do other species, but it also may be found on grassy hillsides, especially in the spring.   Here it forms extensive runways in regions where the grass grows well for a short time.   When the grass dries up, this mouse moves to another location.

Genus LAGURUS
Short-tailed Meadow Mice

Fig. 300.   *Lagurus curtatus*.   Short-tailed Meadow Mouse, or Sagebrush Meadow Mouse.

Figure 300

Total length 115 to 140 mm.; tail 20 to 27 mm.; hind foot 16 to 17 mm.; color of upper parts light buffy gray; ears and nose tinged with buff; underparts and feet pale buff; fur long and lax; mammae 8; plantar tubercles 5; auditory bullae large; third molar with 4 prisms (fig. 301).

These short-tailed mice live in dry localities some distance from water, often occurring among sagebrush.   Little is known of their habits, although they have now been found in nearly all the western states.   They do not seem to make runways, for they usually live where there is not enough grass for runways.   Hundreds of specimens have been taken in Nevada and Washington, but there are only scattered records of this rare mouse in the other western states.

Figure 301

Family MURIDAE
Old World Rats and Mice

These animals, so familiar to everyone, have been introduced into our country, but they are now so abundant that they are more common than any other rodents around human dwellings. They all have tails that are only sparsely haired, and have a very unpleasant odor about them. They are among our most destructive pests, and carry diseases as well.

Key to Genera and Species of European Rats and Mice

1a. Size small, mouse-like; found mainly in buildings and about garbage dumps, but also in the open fields to some extent. Fig. 302. *Mus musculus*. European Mouse, or House Mouse.

Total length about 150 to 175 mm.; tail 60 to 90 mm.; hind foot 15 to 21 mm.; ear 12 to 19 mm.; color grayish brown with the underparts buffy gray, not much lighter than upper parts.

These are the common mice that get into our buildings, and cause us no end of trouble. They may also occur in the open fields in some places, where they build trails that resemble the runways of meadow mice.

Figure 302

1b. Size larger, rat-like; found in buildings, warehouses and about garbage dumps. Fig. 303. *Rattus rattus*. European Rat, Wharf Rat, Roof Rat, Barn Rat, or Black Rat.

Total length 325 to 450 mm. or more; tail 150 to 220 mm.; hind foot 36 to 44 mm.; ear 15 to 21 mm.; color varying; may be all black, brown or grizzled grayish brown, depending upon the subspecies. The common laboratory rats, or white rats, also belong to this species.

These are the most destructive rodents in our country; they are responsible for the loss of many thousands of dollars worth of materials every year, not to mention the effect they have on human life from the transmittance of such diseases as bubonic plague.

Figure 303

A closely related species, *Rattus norvegicus*, now occurs everywhere, especially in cities and garbage dumps. It is the most common rat in our cities. It is larger in body size, but with shorter tail than *Rattus rattus*, and is always brown in color.

Family ZAPODIDAE
Jumping Mice

This family includes only the very long-tailed mice of the genera *Zapus* and *Napeozapus*. They lack the cheek pouches found in the kangaroo rats, and differ from most mice in that they hibernate. The forelegs are small while the hind legs and feet are large. The ears are short, but extend beyond the fur for a short distance. Coloration: upper parts yellowish brown; sides buff or pinkish buff; underparts pure white; tail very long, and tapering to a very slender tip.

Jumping mice build their nests in tall grass on the surface of the ground, generally not far from streams. During the summer they store up a large amount of fat, and in the fall they begin their hibernation in a nest some distance below ground.

Key to the Species of Jumping Mice

1a. **Upper cheek teeth 4, the first very small; tip of tail not white. Zapus** 2

1b. **Upper cheek teeth 3; tip of tail white.   Fig. 304.   *Napeozapus insignis*.   Woodland Jumping Mouse.**

Figure 304

Total length 210 to 250 mm.; tail 125 to 150 mm.; hind foot 28 to 34 mm.; color of upper parts yellowish brown; sides brighter yellowish brown; underparts white.

This mouse lives entirely in the forest, along small streams, and is similar in habits to the western, rather than to the eastern species of Zapus. It does not make runways, but uses those made by other mammals, or else hops about fallen logs and stumps, feeding upon seeds and other plant food.

2a. Size slightly smaller, total length 200 to 220 mm.; found from the Rocky Mountains to the eastern states. Fig. 305. *Zapus hudsonius*.

Tail 119 to 136 mm.; hind foot 28 to 32 mm.; color of upper parts yellowish brown, fur rather coarse; underparts pure white; sides yellowish; tail brown above and white below.

These charming little jumping mice live in the meadows of the northern and eastern states from the Rockies eastward. When one is frightened it may jump for two feet straight into the air, then go bounding off through the grass in erratic leaps. It may jump several feet at a time, perhaps 8 to 10 feet, according to Hamilton. Hibernation is begun in October, and the nest is usually some distance below ground.

Figure 305

2b. Size larger; found west of the Rockies.........................3

3a. Size slightly larger; found along the coastal area and in the Cascade Mountains from British Columbia to California. Fig. 306. *Zapus trinotatus*. Coast Jumping Mouse.

Total length 240 to 260 mm.; tail 140 to 158 mm.; hind foot 31 to 35 mm.; ear 16 to 17 mm.; color of upper parts dusky with a tinge of yellowish; sides rich orange; underparts pure white.

Figure 306

These big fellows live mainly in the foothills and mountains of the coastal area, but may also occur in the valleys in many places. They are always found around water, especially along small streams. They forage along the edge of the stream, sometimes venturing out onto rocks or logs in the water.

3b. Size slightly smaller; found in the Rocky Mountain area from Canada to New Mexico, and in the eastern parts of the Pacific coast states. Fig. 307. *Zapus princeps*. Jumping Mouse.

Figure 307

Total length 222 to 256 mm.; tail 130 to 160 mm.; hind foot 30 to 33 mm.; ear 14 to 16 mm.; color of upper parts yellowish brown to very dark yellowish brown; sides yellowish brown, brighter than the back; underparts pure white.

This mouse lives along streams in the foothills and mountains of the areas east of the Cascades and Sierras; it occurs in valleys only if they are high in elevation.

Family ERETHIZONTIDAE

Porcupines

Fig. 308.   *Erethizon dorsatum*.   Common Porcupine.

Total length 600 to 875 mm.; tail 200 to 260 mm.; hind foot 90 to 110 mm.; ear about 32 mm.; color of upper parts blackish brown, or lighter, depending upon the subspecies.   It is covered all over the back with sharp-pointed quills; tail most heavily armed; underparts buffy.

Figure 308

The porcupine needs no introduction to the majority of people, for its large size and spiny back and tail are well known.   The quills cannot be thrown, but they come out easily, so that if the tail should hit the leg of a person or animal they will stick readily.   Quills have barbs which facilitate their entering the flesh, yet make their removal difficult and painful.   There are no quills on the underside of the animal.   Porcupines live mainly upon juicy plants, but when these are not available they will eat the bark of trees, often climbing to the top of a sapling in order to eat the tender bark.

Family CAPROMYIDAE
Coypu or Nutria

Myocaster coypus. Coypu.

Total length about 1200 mm.; weight reported to be about 50 pounds; size about that of a beaver; habits like those of a muskrat; tail round and almost naked. Color reddish brown; incisors orange; mammae on the back rather than on the belly.

This large rodent has been introduced from South America as a fur-bearing animal, and has been kept in captivity on fur farms. They sometimes escape, and become wild. Small colonies have been established in the wilds in several localities, at least in the states of Washington and Oregon, in British Columbia, and likely in other regions as well. They are reported to do considerable damage to ditches in irrigated regions. Since their fur has little commercial value, they may become a nuisance in time.

Order CARNIVORA

Most people know the carnivores only from the trips to the zoo, but many of them live around us, in spite of their inherent fear of man. Although they do some damage to domestic animals, we would hardly like to see the carnivores exterminated, for they are among the most romantic creatures when it comes to adventure, and among the most valuable when we consider the furs they yield. Carnivores feed almost entirely upon other animals.

Key to the Families of CARNIVORA

1a. Mammals walking on nearly all of foot (plantigrade), 5 toes on each foot 2

1b. Mammals walking on tips of toes (digitigrade); toes either 5 or 4 on front feet, either 5 or 4 on hind feet, but not the same number in any case (either 5-4 or 4-5).............................. 4

2a. Tail short, almost rudimentary; size very large; molars 2 above, 3 below. URSIDAE, page 160.

2b. Tail moderate to long; size smaller; molars 2 above and 2 below, or else 1 above and 2 below.................................. 3

3a. Tail banded with light and dark bands; molars 2 above and 2 below. PROCYONIDAE, page 162.

3b. Tail not banded; molars 1 above and 2 below. MUSTELIDAE, page 163.

4a. Claws blunt and not retractile; rostrum or muzzle long. CANIDAE, page 158.

4b. Claws sharp and retractile; rostrum or muzzle short. FELIDAE, page 170.

Family CANIDAE
Foxes, Coyotes and Wolves

These are familiar animals to everyone, though perhaps not from first hand experience. Since the domestic dog belongs to this family it is hardly necessary to say anything of its general characteristics.

Key to Genera and Species of CANIDAE

1a. Size larger, about that of a medium or large dog; tail less than half the length of head and body; upper incisors lobed. *Canis*...2

1b. Size smaller; tail more than half the length of head and body; upper incisors not lobed.....................................3

2a. Size very large, usually more than 1400 mm. in length; tail more than 375 mm. Fig. 309. *Canis lupus*. Timber Wolf.

Figure 309

Hind foot 180 to 225 mm.; color usually some shade of gray, depending upon the subspecies; fur long and thick in winter, thinner in summer.

The timber wolves were formerly abundant all over America, but with the passing years they have become more and more scarce due to continued hunting and trapping. Some still occur in wilderness areas, but they are most common in Canada.

A closely related species of wolf, *Canis niger*, the red wolf, occurs sparingly in the Southern states, in wilderness areas such as the large swamps, but may also be found in more inhabited regions at times.

2b. Size smaller, usually less than 1200 mm.; tail usually less than 375 mm. Fig. 310. *Canis latrans*. Coyote or Prairie Wolf.

Total length usually 1100 to 1200 mm.; tail about 350 to 395; hind foot about 175 mm.; color usually buffy gray, but shades vary according to subspecies.

Coyotes live in nearly all parts of our country, for they have learned to adapt themselves to the habits of man in such a way that they have actually increased in numbers during the past two decades. Their habits are such that cattle and sheep raisers do not appreciate them, but since they catch so many rabbits and rodents, they may actually be of more benefit than harm if they do not become too numerous.

Figure 310

3a. Tail with dark stripe of stiff hairs down upper side, and no soft hair on underside; muzzle shorter. Fig. 311. *Urocyon cineroargenteus*. Gray Fox.

Total length about 950 to 1000 mm.; tail about 300 to 375 mm.; hind foot about 120 to 125 mm.; color gray, somewhat darker on the back; underparts tawny; feet and tip of tail black.

Foxes are the silent marauders of the farm yards, for they have learned the ways of men so well that they can live right among them without being caught—most of the time. They feed on rodents and other small animals.

Figure 311

3b. Tail without dark strip of stiff hairs down upper side, and soft hair on underside; muzzle longer. *Vulpes*......................4

4a. Size small, less than 900 mm. Fig. 312. *Vulpes macrotis*. Desert Fox or Kit Fox.

Size about 650 to 850 mm.; tail 200 to 300 mm.; weight 3 to 5 pounds; hind foot 95 to 110 mm.; ears unusually large; color of upper parts yellowish gray or grizzled gray; underparts white; tip of tail black.

These little animals live on small rodents and insects. They live in burrows in the ground.

Figure 312

4b. Size larger, more than 900 mm. Fig. 313. *Vulpes fulva*. Red Fox, Black Fox, Cross Fox, or Silver Fox.

Total length about 1000 mm.; tail about 400 mm.; hind foot about 165 mm.; color of upper parts reddish gray; feet and ears black; tip of tail and underparts whitish; colors vary tremendously, however, from nearly all black to silvery gray, and varying shades between. This fox lives over most of the United States, occurring in a number of subspecies.

Figure 313

Family URSIDAE

Bears

Everyone has seen the bears in the zoo, if not out in the woods. One could hardly mistake a bear for anything else—unless he were running too fast in the opposite direction! Bears can still be seen in the wilds in many parts of the country, and they are plentiful in the National Parks such as Yellowstone, Yosemite, and Mount Rainier. Unlike most carnivores, bears are omnivorous; that is, they can eat plant food (mainly berries) as well as flesh. Bears sleep most of the winter, but they do not truly hibernate as do ground squirrels. If their dens are broken into they are wide awake and ready for action.

1a. **Size large, total length 7 feet or more; claws on front feet much longer than those of hind feet; color brown or yellowish brown; cannot climb trees. Fig. 314. *Ursus horribilis*. Grizzly Bear.**

Figure 314

Total length 2000 to 2500 mm.; weight nearly 1000 pounds in very large specimens.

The grizzly bear is scarce in America, occurring only in the Rocky Mountains, and there only occasionally. This bear does not have an amiable disposition, and cannot associate with man in a safe way, even in National Parks. It is comforting to know that they cannot climb trees. The Alaska Brown bear of northern Canada and Alaska is even larger, but it is a closely related species.

1b. **Size smaller, total length less than 7 feet; claws on front feet very little, if any, longer than those of hind feet; color either black, brown, or cinnamon; expert at climbing trees. Fig. 315. *Ursus americanus*. Black Bear, Brown Bear, or Cinnamon Bear.**

Figure 315

Total length 1300 to 1600 mm.; tail 120 mm.; weight up to 500 pounds, but usually less.

These are the common bears of the United States. In general, they are harmless creatures if left alone, for they have good dispositions. A female with cubs, however, should always be given the right of way.

Family PROCYONIDAE
Raccoons, Ringtails

Fig. 316. *Procyon lotor.* **Raccoon.**

Figure 316

Total length about 800 mm.; tail about 250 to 275 mm.; hind foot 120 to 130 mm.; color of upper parts dark grizzled gray, often blackish; underparts light brown; tail alternating black and gray bands.

Raccoons are fascinating creatures of the woods. They feed mainly along streams where they can wash all their food in the water. Here they like to catch crayfish and a wide variety of small animals. Raccoons are generally harmless to man, and may even be considered beneficial because of the rodents and other harmful animals they eat. They are expert climbers. During the summer they store up a large amount of fat to sustain life during their winter's sleep.

Ring-tails

Fig. 317. *Bassariscus astutus.* **Ring-tail.**

Figure 317

Total length about 800 to 900 mm.; tail over 400 mm.; hind foot about 75 mm.; color of upper parts blackish gray; underparts yellowish gray. The tail is very long and bushy, and made up of alternating black and white bands.

The Ring-tails live mainly in the Southwest, where they feed upon a wide variety of small desert animals. They make their dens among rocks or in caves, or in a hole in a tree, generally near water. They are said to be easily tamed, and have been known to make their residence about cabins.

Family MUSTELIDAE
Weasels, Skunks, Martens, and Otters

This family contains some of the most savage mammals of our country. In general, they are slender animals with very powerful muscles, with strong teeth, and an irritable disposition. They are harmful to domestic animals, but provide some of the finest furs we have.

Key to Genera and Species of MUSTELIDAE

1a. Toes fully webbed; teeth 32 or 36.............................2
1b. Toes not fully webbed; teeth 34 or 36.........................3
2a. Size about 1200 mm.; tail about 300 mm.; teeth 32; found in the coastal waters of the Pacific Ocean. Fig. 318. *Enhydra lutris.* Sea Otter.

Figure 318

Hind foot about 150 mm.; color of upper parts brownish black; underparts about the same color.

Sea otters occurred formerly all along the west coast, but have been almost entirely exterminated. In recent years they have been afforded legal protection, so their numbers have increased some, but they are still rare; they now occur sparingly along the central California coast. They feed on shell fish, and spend most of their time in the kelp beds not far from shore, especially about rocky islands. They play together in a most delightful manner, and seem to be especially fond of their young.

2b. Size about 1100 mm.; tail about 400 to 450 mm.; teeth 36. Fig. 319. *Lutra canadensis.* American Otter.

Figure 319

Hind foot about 120 mm.; color of upper parts dark brown; underparts grayish.

Otters are still found in the more wilderness areas of our country, but are very rare. They are playful animals, and have been observed making slides on a muddy stream bank where they will roll and slide for hours, climbing up the bank just to slide down again into the water. They catch small aquatic animals, feeding on very few game fish.

3a. Size 1000 to 1100 mm.; tail 200 to 225 mm.; hind foot nearly 200 mm.; weight over 30 pounds; teeth 38. Fig. 320. *Gulo luscus*. Wolverine.

Figure 320

Color of upper parts dark brown to blackish brown; buffy band along sides.

Wolverines are the most fierce of the weasel tribe, for they fear neither man nor beast. They are the veteran marauders of the trap line; they will steal the animals out of the traps, mutilate them, but not eat them. They spring traps, and cause the trapper so much trouble that he must either kill the wolverine or move to another locality. They are so crafty that it is very difficult to trap them.

3b. Size smaller; teeth 34 or 38....................................4

4a. Body low, wide and heavy; tail ¼ or less of total length; claws of forefeet 30 mm. or more in length; teeth 34. Fig. 321. *Taxidea taxus*. Badger.

Figure 321

Total length 700 to 825 mm.; tail 125 to 175 mm.; hind foot about 65 mm.; color of upper parts grizzled brownish gray; black markings on top of head, back of ears, cheek, feet, and legs; white stripe from nose to back of neck, and on down the back varying distances. The animal has long guard hairs, much longer on the sides of the body than on the back, which accentuates his low, squat, appearance.

Badgers are fierce animals of the open country, where they live among ground squirrel or prairie dog colonies. Here they dig out their victims by using their powerful legs and long claws. The animal is so strong that no dog will dare dig one out— at least the second time. They are adept at digging, and once a badger is down in the ground it is almost impossible to dig him out, so fast he burrows and so well does he close up the hole behind him. After eating his fill of squirrels, rabbits, or the like, he holes up for several days. In the winter the badger may stay underground for a prolonged period.

4b. Body not low and wide, and not very heavy; tail longer than ¼ of total length; claws much shorter; teeth 34 or 38..................5

5a. Color pattern black and white................................6

5b. Colors mainly brown with little or no white (except in winter when the entire animal may be white with black tip on tail).... 9

6a. Muzzle long and proboscis-like; color black except for a white tail and a wide white band down center of back. **Fig. 322.** *Conepatus mesoleucus.* **Hog-nosed Skunk.**

Figure 322

Total length about 650 to 700 mm.; tail about 300 mm.; hind foot about 75 mm.

The Hog-nosed Skunk is distinctly a southwestern animal, but his habits are similar to the other skunks. He feeds upon insects more than anything else, but includes in his diet small rodents, birds and eggs, and even the fruit of cactus. Skunks have very few natural enemies because of their effective protective device, but horned owls and even coyotes and bobcats will sometimes catch skunks.

6b. Muzzle shorter, not proboscis-like; color black, with white areas differing from those of *Conepatus*............................7

7a. White areas on back a series of 4 interrupted white stripes, but usually appearing to be a series of spots; size decidedly smaller than Mephitis, the striped skunks; found from the Atlantic to the Pacific coast (see map). Fig. 323. *Spilogale putorius*. Spotted Skunk.

Figure 323

The little spotted skunk (often known as civet cat) lives in waste lands, in cultivated fields, around old buildings, and in woodsy and weedy places in general all over the country, but is more common in warmer regions than in the colder, northern areas. While many people dislike these animals they are valuable destroyers of rodents and insects, and if left alone they will cause no damage to man at all. They will enter chicken houses and eat eggs and young chickens, so it is wise to make chicken houses "skunk-proof," for skunks do far more good than harm, and should not be killed.

7b. White area 2 white stripes down the back. Mephitis............8

8a. Back black, or with white areas containing no black hairs; auditory bullae small. Fig. 324. *Mephitis mephitis*. Striped Skunk.

Total length 600 to 800 mm.; tail 200 to 315 mm.; hind foot 60 to 84 mm.

Figure 324

Striped skunks are not the offensive creatures many people think them to be. It is true that they are able, when sufficiently aroused, to release an extremely ill-smelling musk which has somewhat lasting qualities. This liquid is released from two glands, one on each side of the anus. These animals are good natured, preferring to go about their business of foraging for insects and mice. They live in holes not far from water, generally excavated for them by ground squirrels or badgers. It is a common sight to see Mrs. Skunk followed by her four to eight or more offspring, trotting along a ditch bank at dusk. The male does not as a rule live with the young family. The fur is extensively used, those pelts having the least white being the most valuable. Young skunks make affectionate and interesting pets.

8b. Back black or with white areas containing black hairs mixed with the white; auditory bullae large; found only in southern Arizona, southern New Mexico, southern Texas, and south throughout Mexico. *Mephitis macroura.* **Hooded Skunk.**

Total length, 558 to 790 mm.; tail, 275 to 435 mm.; hind foot 58 to 73 mm. When the back of the skunk is white this species can be easily distinguished from the other by the black hairs mixed with the white hairs. If the skunk has a black back (which does happen now and then) the white stripes will be wide apart in this species, even out along the sides of the body, while in the common skunk the white stripes will be close together down the center of the back. The odor is the same in both species—and a little goes a long ways!

9a. Tail long and bushy, more than one-third of total length; premolars 4 above and 4 below; size rather large; teeth 38. *Martes*..10

9b. Tail shorter, not bushy, less than one-third of total length; premolars 3 above and 3 below; teeth 34; size smaller. *Mustela*...11

10a. Size large, about 975 to 1050 mm.; tail about 400 mm.; hind foot 128 to 138 mm.; color grayish brown; underparts blackish, no yellow area on throat or breast. Fig. 325. *Martes pennanti.* **Fisher.**

Figure 325

This valuable animal is almost as large as a fox, with short legs, and a rather short tail—at least shorter than that of a fox. It is rare because of over-hunting and over-trapping. The fisher is a wanderer traveling through timber (it is an agile climber), swimming across lakes or over northern marshes. It nests in a cavity in a tree or log or in a rocky cave. In spite of its name fish do not make up much of the fisher's food supply; it preys on porcupines, foxes, skunks, rabbits, and mice.

10b. Size smaller, about 600 to 650 mm.; tail 200 to 235 mm.; hind foot 75 to 100 mm.; weight 1½ to 2½ pounds; color yellowish brown; throat and breast orange. **Fig. 326.** *Martes americana.* **Marten.**

This animal is perfectly at home in the trees of the northern woods, where it hunts squirrels and chipmunks with great agility. It is not so rare as is the fisher, yet it is only relatively common in woods not much invaded by man. It is of a curious nature and is an easy prey for the trapper, who highly prizes its valuable pelt.

Figure 326

11a. Size rather large, about 500 to 600 mm.......................12

11b. Size smaller, less than 450 mm............................13

12a. Toes webbed; soles of feet naked; tail nearly 200 mm.; hind foot about 70 mm.; color dark chestnut brown; underparts somewhat paler; lives mainly about water. **Fig. 327.** *Mustela vison.* **Mink.**

The mink is one of our most common fur-bearing mammals, although over-trapping has seriously depleted its numbers. It may be found in many parts of the country where it lives mainly along streams, wandering over its range in search of muskrats and aquatic animals such as fish, frogs, and insects. It makes its nest often under tangled tree roots along secluded streams.

Figure 327

12b. Toes not webbed; soles of feet hairy; total length about 500 mm.; tail only about 100 mm.; hind foot about 55 mm.; color light buff with a dark area in the middle of the back; underparts light buff;

black band across face, including eyes; tail with black tip; feet and lower legs black. **Fig. 328.** *Mustela nigripes.* **Black-footed Ferret.**

This, our largest weasel, lives near ground squirrel and prairie dog colonies, and feeds almost entirely upon these animals. He is therefore a beneficial animal in keeping these large rodents under control. Like all weasels, he is sly and cunning, and tireless in his hunting.

Figure 328

13a. Tail long, about 125 to 175 mm., more than 47 percent of the length of head and body; underparts usually yellowish; body size large, much larger than 18a and 18b. **Fig. 329.** *Mustela frenata.* **Long-tailed Weasel.**

Total length varying greatly depending upon the subspecies (there are 41 of them), but from about 300 to 435 mm.; hind foot 39 to 55 mm.; color light brown with a yellowish belly; tail with a black tip. Many subspecies turn white in winter, but those in warmer climates do not. Found in nearly every part of the United States and southern Canada.

Figure 329

Anthony believes there is a special type of weasel for every ecological niche—but all weasels share in possessing amazing agility, tirelessness, and fearlessness. They may be seen any hour of the day or night, summer or winter. Any small mammal is their prey, not excluding some larger than they are. Males are on the average about twice as heavy as the females.

13b. Tail short, from 15 to 100 mm.; less than 47 percent of the length of head and body; underparts white or partly white; body size small, much smaller than that of 17a; color white in winter in northern localities 14

14a. Tail very short, only 20 to 35 mm.; hind foot only 18 to 25 mm.; total length 150 to 200 mm.; chin, throat, and lower belly white; middle part of belly brown; upper parts brown. Fig. 330. *Mustela rixosa.* Least Weasel.

Figure 330

This is the smallest carnivore in America, but is as fierce as any weasel. It lives either in the forests or in cultivated fields and waste lands where it is in almost constant pursuit of some small rodent or other tiny creature.

14b. Tail longer, about 45 to 100 mm.; total length 188 to 328 mm.; hind foot 23 to 44 mm.; upper parts brown; all of underparts white. Fig. 331. *Mustela erminea.* Ermine.

Figure 331

This weasel is the ermine which sometimes has the misfortune to fall into the trap and thus makes its way into the ladies' coats. In summer it is a small inconspicuous brown creature with a white belly, but in winter it is pure white except for the black tip of its tail.

Family FELIDAE
Cats

General characteristics of this family are well known to everyone who has had a house cat, or who has admired at the zoo such famous members of the family as lions, tigers, cougars, panthers and bobcats.

Key to Genera and Species of FELIDAE

1a. Tail very short, only 100 to 180 mm. *Lynx*......................2

1b. Tail longer, 375 to 900 mm. *Felis*............................3

2a. Tail only about 100 mm.; feet very large, about 200 to 225 mm.;
total length 825 to 950 mm.; color gray, sprinkled with brown, with
tip of tail solid black; ear tufts very
long, about 37 mm.; long side whiskers on face; fur long. Fig. 332. *Lynx
canadensis.* Canada Lynx.

This bobcat is found in the northern
mountains, and never comes down
into the United States very far. Its
big feet, long whiskers, and short tail
will identify it. The animal lives
largely on the snow shoe rabbit,
squirrels, skunks, and foxes.

Figure 332

2b. Tail 170 to 180 mm.; feet smaller, about 180 to 200 mm.; total
length 800 to 900 mm.; color reddish with tip of tail banded with
light and black areas; ear tufts short, only about 15 mm.; side
whiskers short; fur short.   Fig. 333.   *Lynx rufus.*   Bobcat; Bay
Lynx.

This bobcat, resembling an over-
grown house cat, is the common one
all over the United States. In spite
of heavy trapping the bobcat still continues to hold its own. It does some
harm to domestic animals, but does
far more good than harm in keeping
down rodents and rabbits. They use
rocky caves, hollow logs, or dense
thickets for their dens.

Figure 333

4a. Size large, total length 1800 to 2400 mm.; tail 750 to 900 mm.; weight over 100 pounds. Fig. 334. *Felis concolor*. Mountain Lion, Cougar, Panther, Puma, or Catamount.

Color yellowish brown, dark stripe down center of back, becoming black on the tail; underparts lighter.

The mountain lion is never common, but still occurs in many places in the western states, and in some of the wilderness areas of the east. Most people fear this animal, but actually there are only two or three authentic records of a mountain lion killing a person. They do considerable damage to livestock, but if they are not allowed to become too common they are actually beneficial, for they keep the other animals in a natural balance. Where they have been almost entirely killed off, deer and rabbits have become so abundant that it was better to have had more mountain lions. They are wide ranging hunters, covering a territory of 30 to 50 miles.

Figure 334

4b. Size smaller, total length about 1000 to 1100 mm.; tail about 375 mm. Fig. 335. *Felis yagouaroundi*. Jaguarundi, or Eyra.

This little cat lives in the Southwest and in Mexico. It is found in two phases: the gray phase, in which it has a grizzled salt and pepper appearance, and a red phase. It inhabits dense thickets, where underbrush is too impenetrable for a larger animal, but where it can capture rodents, birds, and rabbits. It is an agile tree climber, and may be seen during the day as well as the night.

Figure 335

5a. Size large, over 2000 mm.; tail about 600 mm.; color tawny with many black spots. Fig. 336. *Felis onca*. Jaguar.

Figure 336

This big cat is one of the most savage in America; fortunately, it does not extend far into the United States, occurring sparingly in the southern parts of Arizona in the remote mountain ranges near the Mexican border. It ranges far in its hunting, one animal utilizing a huge territory. Anthony (1928) reports it from as far north as Colorado and central California. He also reports it from southern Texas and southern New Mexico, although the animal is certainly very rare within the boundaries of the United States. These northern jaguars are smaller and less aggressive than the South American forms, but they are still fearsome beasts, and one might well keep his distance if he should have the rare opportunity to see one of these beautiful creatures.

5b. Size smaller, about 1200 mm.; tail about 375 mm.; color tawny with large black spots and stripes. Fig. 337. *Felis pardalis*. Ocelot or Tiger Cat.

Figure 337

This animal may appear to look somewhat like a jaguar, but it is much smaller, and much less aggressive. It occurs sparingly in the wilderness regions along the Mexican border, but it is evidently much less common than it was in the past. Nelson (1930) says it was rather common in dense thorny chaparral in the lower Rio Grande valley, but its abundance has steadily decreased. While most people fear the larger members of the cat family, they are among our most picturesque wild mammals, and it is with regret that we must admit that these big cats will soon be gone from the United States.

A closely related species, *Felis wiedii*, lives in much the same area as the ocelot, but it is much smaller. It is spotted in much the same manner, and most people have always considered this species a young ocelot. The common name for this beautiful cat is Margay. It makes an excellent pet, and will keep all the neighbors dogs in their own yards.

Order PINNEPEDIA
Seals and Sea Lions

These animals are difficult to identify accurately because ordinarily it is impossible to see the whole animal. If one is fortunate enough to find them out of water on a rocky island, then identification should be easy. If they are seen as they bob up above the water for an instant, you will have a hard time getting results by these keys.

Key to Genera and Species of Seals and Sea Lions

1a. Size large, 8 feet or more in length..........................2

1b. Size smaller, 5 to 6 feet in length............................7

2a. Fleshy growth over nose.....................................3

2b. No fleshy growth on nose....................................4

3a. Fleshy growth hanging down like the trunk of an elephant on the male. Fig. 338. *Mirounga angustirostris*. Elephant Seal.

Figure 338

Size 12 to 18 feet in length; found only off the coast of extreme southern California, where a herd frequents Guadalupe Island.

These huge seals are now nearly extinct, and do not occur in the United States at all, although they formerly occurred along the southern California coast. The large trunk-like protuberance above the mouth gives them the name of Elephant Seal.

3b. Fleshy growth above tip of nose. Fig. 339. *Cystophora cristata.* Hooded Seal.

Size about 8 feet; found from northern New England to Greenland, where in small groups it frequents ice hummocks to bear its young.

This is one of the rather common seals of the north Atlantic, and chooses icy cliffs on artic islands for its breeding grounds. It does not occur in large colonies, but prefers to remain in small groups.

Figure 339

4a. Males with huge tusks extending from upper jaw. Fig. 340. *Odobenus rosmarus.* Walrus.

Size 10 to 12 feet; weight 2000 to 3000 pounds; found in the arctic. Shrimp, starfish, and molluscs are its principal food.

Walruses do not migrate south in winter, but remain in the arctic seas the entire time. They make an imposing sight with the huge tusks and heavy necks of the males. Eskimos have hunted these animals for many years as a staple food, and for their skins and tusks, and now their numbers have been greatly reduced until they are becoming scarce. Two subspecies occur, one on the Atlantic, the other on the Pacific side of the Arctic Ocean.

Figure 340

4b. No tusks present **5**

5a. Color brown to blackish brown; found only on the Pacific coast.

(1) Size 12 to 15 feet, color brown, weight 1000 to 1800 pounds. *Zalophus jubata*. Northern Sea Lion.

(2) Size 8 to 10 feet, color blackish brown, weight 500 to 900 pounds. Fig. **341**. *Eumetopias californianus*. California Sea Lion.

These animals are a common sight on rocky islands and shores; both species often occur together.

Figure 341

5b. Color plain gray; found in the north Atlantic or the north Pacific..6

6a. Long flattened bristles extend downward around mouth giving a bearded appearance; found from Newfoundland north to the coast of Alaska. Fig. 342. *Erignathus barbatus*. Bearded Seal.

Size 10 to 12 feet.

• Bearded Seals occur only in Arctic seas, and form one of the chief foods of the Eskimos. Although some seals will break holes in the ice for fishing, this one does not, but takes advantage of cracks already present for its fishing operations.

Figure 342

6b. No unusually long bristles about mouth; obscure black spots on underparts; found from Nova Scotia to Greenland. Fig. 343. *Halichoerus grypus*. Gray Seal.

Size 10 to 12 feet.

The Gray Seal is one of the uncommon seals along the American coasts, but it may sometimes be found in rocky pools where the tide is rushing in and out among caves, swirling and boiling until it would seem that the animal would be dashed to pieces on the rocks. Males do much fighting during the breeding season, and may be badly scarred for their bouts.

Figure 343

7a. Color pattern contrasting brown and white, even a[ ] and white at times............................

7b. No contrasting colors.................................

8a. Color dark brown with yellow band around neck, forelimbs and posterior region. Fig. 344. *Phoca fasciata.* Ribbon Seal.

Figure 344

Size about 5 feet; found in the Aleutian Islands and along the Alaska coast.

The Ribbon Seal is one of the most brightly marked seals along American coasts. The males show the most contrast; the females have the light "ribbons" more obscure, but they are plainly visible. These are not common seals, and they never come to the coastal part of the western states, remaining continuously in Alaskan waters.

8b. Color yellowish with dark brown area on head and on neck. Fig. 345. *Phoca groenlandica.* Greenland Seal or Harp Seal.

Figure 345

Size about 6 feet.

This is one of the most important seals commercially, and one of the common seals along the Atlantic coast. They appear off Newfoundland in September, and migrate as far south as the Gulf of St. Lawrence where they spend the winter. In early spring the young are born on cakes of ice in the Gulf or a little farther north, and remain on these ice pans until old enough to take to the water.

. Upper parts mainly plain black, not spotted; shoulder grayish; face brown; underparts reddish brown; fur thick and glossy. **Fig. 346.** *Callorhinus ursinus.* **Northern Fur Seal.**

Figure 346

Size about 6 feet; found from Alaska south to California.

Fur seals were among the most abundant of seals many years ago, but intensive hunting nearly exterminated them until the United States government began to protect them. Now they are under controlled hunting, and a regular fur crop is harvested each year. The population now remains rather large. They are seldom seen along the west coast in winter for they remain well off shore.

9b. Color varying, but always spotted..........................10

10a. Upper parts brown or blackish brown with faint light yellowish rings. Fig. 347. · *Phoca hispida.* **Ringed Seal.**

Figure 347

Size about 6 feet; found from Labrador north to the Bering Sea.

While the markings of this seal are rather characteristic, the extra long first digit of the fore-flipper is also a good mark (if you get that close). They occur in arctic seas only, but are circumpolar in distribution.

10b. Upper parts varying from yellowish gray with brown spots to dark brown with yellowish spots; young are white. **Fig. 148.** *Phoca vitulina.* Harbor Seal.

Figure 348

Size about 5 feet.

These are the common little spotted seals that may be seen almost anywhere along either the Atlantic or the Pacific coasts. They are enough to make one smile as they bob their clean shaved little heads up out of the water to take a quick look at you in your boat. First you see one here, then one there, now in sight, now gone, until you wonder how many there are around, and why they keep bobbing up and down. You can spend an interesting hour watching them—if they will stay around you that long.

Order EDENTATA
Armadillo

Fig. 349. *Dasypus novemcinctus.* Armadillo.

Total length about 800 mm.; tail about 375 mm.; hind foot 100 mm.; weight 9 to 15 pounds; body covered with a series of circular horny bands making a shell around the entire animal;

Figure 349

ears prominently projecting above back of head; snout long and slender; teeth very small and poorly developed; claws long and sharp.

The armadillo lives mainly in the deserts of Texas, but in recent years has spread across the South to Florida. It feeds on a host of small soft insects and worms. It is harmless and continues to survive, although it is killed frequently.

Order CETACEA
Whales, Porpoises, and Dolphins

This group of mammals may not appear to be true mammals to the casual observer, for they do have a superficial resemblance to fish. However, they are furred animals with mammary glands and most of the other mammal characteristics, so a closer look is called for.

When it comes to identifying the various kinds of whales and porpoises we have another problem on our hands. In the first place, it is seldom that we see these creatures, and when we do, it is a fleeting glimpse of a dorsal fin as the animal comes up for a breath of air. If one watches carefully, the whale is likely to come up several times. You may see some characteristics—but we do not promise too much! Rarely a whale may be washed in dead on the beach—if so, there's your chance!

Key to Genera and Species of Whales, Porpoises and Dolphins

1a. Size 15 to 100 feet in length; two nostrils (or blow holes) present (except in *Berardius*, 4a, and in *Physeter*, 5a)..................2

1b. Size 5 to 12 feet in length; only one nostril present. Small whales, porpoises and dolphins......................................15

2a. Size 35 to 100 feet...3

2b. Size 15 to 30 feet..10

3a. Size 35 to 40 feet...4

3b. Size 50 to 100 feet..5

4a. Head beak-like; only one nostril present; dorsal fin small, and placed on the posterior part of the back; color all black except for the grayish or whitish area on the lower belly; head small with slender, beak-like jaws, and with one or two large teeth near the tip of each lower jaw. Fig. 350. *Berardius bairdii*. Baird Whale.

Figure 350

This is a very rare whale known only from a few specimens which have been washed onto shores from the Bering Sea south to California. No observations appear to have been made on living specimens, possibly because of its scarcity.

4b. Head not beaked; two nostrils present; no dorsal fin present at
all; color light gray, somewhat mottled, but may be almost black,
with the mottling mainly on the upper surface; teeth more than
2, but entirely absent in adults; 2 or 3 longitudinal grooves present on throat; eye above corner of mouth; length about 40 feet,
but may be as long as 44 feet. Fig. 351. *Eschrichtius gibbosus.*
Gray Whale.

These whales spend the winter
months along the coast of southern
California. Here the young are born
in early spring in shallow bays.
Later in the spring they migrate north
all along the coast as far as the
Bering Sea, and the Arctic Ocean.

Figure 351

5a. Lower jaws very narrow; upper jaws extremely heavy; only one
nostril present; size 70 to 85 feet; 40 to 50 teeth in the lower jaws,
no teeth in upper jaws; eyes well above corners of mouth; no
dorsal fin present; color blackish above but rather grayish underneath. Fig. 352. *Physeter catodon.* Sperm Whale.

This giant whale with the enormous head is found in all seas that
do not freeze over in winter, but it
has been hunted so extensively that
its numbers are greatly reduced,
nearly to the point of extinction. This
is the most valuable of all whales
commercially.

Figure 352

5b. Jaws not as in the Sperm Whale; two nostrils present.........6

6a. Small dorsal fin present....................................7

6b. No dorsal fin present.......................................9

7a. Color black or blackish brown; underparts whitish.............8

7b. Color gray to bluish gray, somewhat mottled; underparts white to
yellowish; size 60 to 100 feet, largest of all whales; head rather
flat on top; dorsal fin and pectoral fins small for the size of the
whale; eyes near corner and below level of mouth; throat and belly
with longitudinal ridges and folds. Fig. 353. *Sibbaldus musculus.*
Sulphur-bottom Whale.

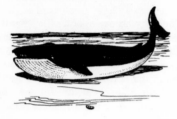

This giant has been known to reach
the length of 105 feet and to weigh
up to 150 tons. It occurs in both Atlantic and Pacific waters, and seems
to avoid the tropics. Since it lives
in very deep water, and is a powerful swimmer, it is seldom captured
by the whalers.

Figure 353

8a. Pectoral fin very long and large, scalloped on the lower margin;
body short and thick; lower jaw projects beyond upper jaw; eye
near corner of mouth, and well below level of mouth; dorsal fin
low and humped; knob-like protuberances scattered over back; longitudinal ridges and folds on throat and belly; size about 50 feet;
color back mottled with white on the underparts. Fig. 354. *Megap-
tera novaeangliae.* Humpback Whale.

These whales occur in all oceans,
but are not common anywhere. Their
peculiar humped appearance makes
them easy to distinguish even at a
distance. They do not yield as much
oil or whalebone as do most other
whales, and are not very important
commercially.

Figure 354

8b. Pectoral fin short; body slender and long, 65 to 80 feet; small fin-like dorsal projection on the posterior part of the back not far from the tail; many deep grooves on throat and belly; eyes near corner of mouth; color black to blackish brown with a white belly. Fig. 355. *Balaenoptera physalus.* Finback Whale.

Figure 355

There are several species of this whale, but the one above is the common one. It occurs in both Atlantic and Pacific waters. *Balaenoptera borealis* is smaller, only about 50 feet long, but is rare in American waters. It was abundant years ago along the Pacific coast. *Balaenoptera davidsoni* is only 30 to 33 feet long, and occurs along the Pacific coast. *Balaenoptera acutorostrata* has the body relatively more robust than does *Balaenoptera physalus*, but it is only 20 to 30 feet in length, and occurs in the North Atlantic from New Jersey northward.

9a. Head more than one-third of total length; upper jaw highly arched; size 50 to 65 feet; color black to blackish brown. Found in the North Pacific, North Atlantic and Arctic seas. Fig. 356. *Balaena mysticetus.* Bowhead, or Greenland Right Whale.

This large whale, and the two species just below, are among the most valuable oil-producing whales of the northern seas.

Figure 356

9b. Head less than one-third of total length; upper jaw not so highly arched; teeth absent, but two large slabs of whalebone occur in the mouth, the edges of which are fringed for straining small floating animals of the ocean; dorsal fin absent; no grooves or folds on throat; color black, or black mottled with white; size 50 to 70 feet. Genus Eubalaena.

Figure 357

Found in Pacific waters from Lower California to Alaska, but more common from Oregon or Washington northward. Numbers have been greatly reduced in the past years by extensive whaling operations. Fig. 357. *Eubalaena sieboldii.* Pacific Right Whale.

Found in Atlantic waters from South Carolina northward, but more common farther north. *Eubalaena glacialis.* North Atlantic Right Whale.

10a. Dorsal fin small, near the tail; prominent beak present; teeth only 2 11

10b. Dorsal fin large, near the head; beak, if present, very small; teeth more than 2 13

11a. Size 24 to 30 feet in adults; body rather heavy with the forehead bulging; well developed beak present; color varies from black to

Figure 358

light brown or almost yellow, whitish about the head, and lighter on the underside; only two small teeth are present, one on each side at the tip of the lower jaw, neither of which is plainly visible. Found in the Arctic and the North Atlantic waters south as far as New York. Fig. 358. *Hyperoodon ampullatus.* Bottle-nosed Whale.

11b. Size smaller 12

12a. Beak longer; size 12 to 22 feet; color black, mottled on the belly.
Size about 16 feet; found in the North Atlantic. *Mesoplodon
bidens.* Sowerby's Whale.
Size about 12 feet; found along the coast of Massachusetts to
New Jersey. *Mesoplodon densirostris.* Atlantic Beaked Whale.

Size about 22 feet; found in the North
Atlantic. *Mesoplodon gervaisi.* Ger-
vais' Beaked Whale.
Size about 16 feet; found from the
Bering Sea to Oregon. *Mesoplodon
stejnegeri.* Stejneger's Beaked Whale.
Size about 16 feet; found along the

Figure 359

coast of south Atlantic states. Fig.
359. *Mesoplodon mirus.* True's
Beaked Whale.

12b. Beak shorter; size 18 to 20 feet in length; color varies from black

to gray or even white, especially on
the head and even back as far as
the dorsal fin; underparts lighter;
only two teeth are present, and these
are located at the tip of the lower
jaw. Found in both Atlantic and
Pacific waters. Fig. 360. *Ziphius
cavirostris.* Goose-beaked Whale.

Figure 360

13a. Profile of head slopes gradually to tip of snout; color black or
black and white; size 16 to 22 feet; pectoral fin short..........14

13b. Profile does not slope gradually to snout, but protrudes over the
tip of upper jaw; color always black; size 15 to 19 feet; pectoral
fin long.

Size 15 to 19 feet; pectoral fin longer
than one-sixth total length; found
along the Atlantic coast south to New
Jersey. Fig. 361. *Globicephala me-
laena.* Atlantic Blackfish.
Size about 15 feet; pectoral fin shorter, less than one-sixth of total length;
found from New Jersey to the Gulf of

Figure 361

Mexico. *Globicephala macrorhyncha.*
Short-finned Blackfish.
Size about 15 feet; found in the north Pacific Ocean. *Globice-
phala scammonii.* Pacific Blackfish.

14a. Color all black; body more slender than in the killer whales
(*Grampus*); total length about 16 feet; the pectoral fins are only

about one-tenth as long as the body;
dorsal fin short and pointed backward, not nearly so long as in the
killer whales; teeth 8 to 11 in each
jaw. This whale is found in all seas,
and feeds mainly upon squids. Fig.
362. *Pseudorca crassidens*. **False
Killer Whale.**

Figure 362

14b. Color black and white; size 20 feet. Found in all seas. **Fig. 363.**
Grampus orca. Atlantic Killer Whale.

Found along the Pacific coast.
Grampus rectipinna. **Pacific Killer
Whale.**

The Killer Whale is aptly termed
the wolf of the seas. Travelling in
packs, they attack the largest whales,
tearing and slashing with their sharp
teeth until the victim is hacked to
pieces.

Figure 363

15a. Long tusk (up to 9 feet) present in males; females usually do not
have the tusk exposed externally, although a vestige may be pre-

sent; total length about 12 feet; color
dark gray on the back; mottled light
and dark gray on the sides, and
white underneath. Found in Arctic
seas on both sides of the continent.
Fig. 364. *Monodon monoceros*. **Narwhal.**

The tusk in the male Narwhal is
actually a twisted tooth which is
greatly developed as a tusk. In the
opposite jaw is another tooth which

Figure 364

is also twisted, but never extends beyond the jaw. The teeth of the
female are like the non-developed tooth of the male, except in unusual cases where the tooth may actually protrude.

15b. No long tusk 16

16a. Jaws protrude, giving a beak-like appearance. Porpoises....19

16b. No beak present 17

17a. Color all white; size 11 to 12 feet; no dorsal fin present; middle
of back well arched; head rounded; pectoral fins short and broad; teeth rather far apart, small and pointed, only about 9 on each side of one jaw. Found in Arctic and subarctic seas south at least as far as Massachusetts. Fig. 365. *Delphinapterus leucas.* White Whale.

Figure 365

17b. Color not white...18

18a. Dorsal fin nearer tail; mouth much below tip of snout, snout short and thick, rounded; dorsal fin low but evident, curved backward; the single nostril is far back on the head; color black above and light beneath; size 9 to 13 feet in length. It is found in all seas, but seems to be rather rare everywhere. It has been taken along the Atlantic coast in scattered localities. Fig. 366. *Kogia breviceps.* Pigmy Sperm Whale.

Figure 366

18b. Dorsal fin nearer head; mouth at tip of snout; teeth 4 to 6 in each jaw; color varies from nearly all black, dark gray with white head or white anterior areas, or mottled gray and white; size about 10 feet. Found in both North Atlantic and North Pacific Oceans as far south as California and New Jersey. Fig. 367. *Grampidelphis griseus.* Grampus.

The actions of this whale are very similar to those of the killer whales, although it feeds upon fish and crustaceans rather than upon other whales or large fish.

Figure 367

19a. Beak very long...20

19b. Beak short...26

20a. Color of upper parts gray or greenish gray...................21

20b. Color of upper parts black................................22

21a. Color gray to greenish gray, sometimes spotted; size 9½ feet. Found along the Atlantic coast from Maine to Texas. Fig. 368. *Tursiops truncatus.* Bottlenosed Dolphin.

Figure 368

A similar species, *Tursiops gillii,* the Cowfish, occurs along the Pacific coast, at least in California. It is slender, with the snout somewhat beak-like with the dorsal fin in the middle of the back; the teeth are 48 in the upper jaws and 46 in the lower; color black with the underparts somewhat lighter.

21b. Color gray on the sides, head black, back, fins, and tail greenish black, belly and throat white; body slender with long narrow jaws forming a beak; each jaw with a row of 80 to 120 conical teeth; pectoral fins narrow; dorsal fin triangular; size 6 to 7½ feet; weight 100 to 160 pounds. Found along the Atlantic coast. Fig. 369. *Delphinus delphis.* Atlantic Dolphin.

Figure 369

A similar species, *Delphinus bairdii,* the Pacific Dolphin, occurs from southern Vancouver Island south to Baja California along the Pacific coast.

22a. Color black above, white below...........................**23**

22b. Color purplish gray to black with numerous white spots, especially on the sides; size about 7 feet; teeth 39 pairs in each jaw, smooth; dorsal fin high and curved backward; pectoral fins broad at the base; snout beak-like and rather stout. Found along the Atlantic coast from Cape Hatteras to the Gulf of Mexico. Fig. 370. *Prodelphinus plagiodon.* Spotted Dolphin.

Figure 370

This dolphin differs from the Common Dolphin *(Delphinus)* by its spotted appearance, and also in having a flat palate without deep grooves inside the tooth rows (as occurs in *Delphinus*).

23a. Found along the Atlantic coast..............................**24**

23b. Found along the Pacific coast..............................**25**

24a. Size about 8 to 8½ feet; beak longer than in the Common Dolphin

(Delphinus); color black or very dark gray with the underparts white; dark longitudinal bands may be present on the belly; body slender; teeth 20 to 27, furrowed and heavier in weight than in the Common Dolphin. Found ordinarily in the Indian Ocean, but also from Virginia to Florida. Fig. 371. *Steno bredanensis.* Rough-toothed Porpoise.

Figure 371

24b. Size about 7½ feet. (See Fig. 369). *Delphinus delphis.* Common Dolphin.

25a. Size about 4 feet; light area on belly very small, color otherwise

black; body slender; no dorsal fin present; jaws forming a definite snout. Occurs in the North Pacific Ocean off the shores of the western states, but it is impossible to say whether or not it is common, for there are almost no records regarding this animal. It is probably rather rare. Fig. 372. *Lissodelphis borealis.* Pacific Right Whale Porpoise.

Figure 372

25b. Size 6 to 7½ feet; light area on belly extensive in size. See fig. 361.) *Delphinus delphis.* Common Dolphin.

26a. Sides with two areas of light color separated by irregular dark

bands; size about 8 feet in length; dorsal fin in about center of back, and very well developed, high and sickle-like; pectoral fins broad at base and pointed; body stout with the forehead sloping; beak merely a rim; teeth 27 on each side of both upper and lower jaws.

Found in the North Atlantic from Cape Cod northward. Fig. **373.** *Lagenorhynchus acutus.* Atlantic White-sided Dolphin.

Figure 373

Found in the North Atlantic north of the United States. *Lagenorhynchus albirostris.* White-beaked Dolphin.

Found in the North Pacific and from Monterey, California north to Puget Sound. *Lagenorhynchus obliquidens.* Pacific White-sided Dolphin.

26b. Sides not having two light areas separated by dark bands....27

27a. Light area of belly does not extend farther forward than the plane of the dorsal fin; color black above, white on belly; size 6 feet;

lower jaws somewhat protruding; teeth small, 23 above and 27 below, alternating with a secondary set of gum teeth which serve as teeth also. Occurs along the coast from Alaska south to California, but rare along the shores of the western states, more common in Alaska. Fig. 374. *Phocoenoides dalli.* Dall's Porpoise.

Figure 374

27b. Light area of belly extends from head to tail; color black above, lighter below, or even white; size usually less than 6 feet; body

short and thick; head short; dorsal fin small and triangular, placed in the center of the back; teeth about 20 to 26 above, and 24 below. Occurs along both coasts, even ascending rivers. Fig. 375. *Phocaena phocaena.* Atlantic Harbor Porpoise.

Figure 375

These small porpoises do not play in large schools as do the larger species of porpoises, but seem to stay in groups of 2 to 6 or 8. They come to the surface at intervals, but remain partly above the water for only an instant, then disappear again. They are sometimes taken in nets by fishermen.

The Pacific Harbor Porpoise, *Phocaena vomerina,* is found in bays and the open ocean from Alaska to Mexico, and is the common porpoise seen along the west coast.

Order SIRENIA
Manatees and Sea Cows

Fig. 376. *Trichechus manatus.* Manatee, or Florida Manatee.

Figure 376

These are large aquatic mammals with a short wide head, no hind limbs, but a tail-like flipper which is very wide and flat, and with the front limbs rather small, flipper-like. Bones of the body are large and heavy. The sea cows live in warm semi-tropical waters where they feed on sea weeds. They usually remain in shallow waters near the coast. They are rare animals, but a few occur along the Florida coast.

Order PRIMATA

Primates do not occur in the United States as wild animals. Man, of course, is a member of this order; other examples may be seen in zoos. Lemurs, monkeys and apes are the common members of this order. The wild primates are native only in tropical countries.

Order PERISSODACTYLA

This order, although a large one, is not represented by any of the present wild animals of the United States. During the Glacial Epoch the horse group with many other of its relatives roamed the plains of America in large herds. Now, they occur only in other parts of the world. Members of this order include the rhinoceros, tapir, zebra, and domestic animals like horses, donkeys and burros.

Order ARTIODACTYLA

This order is one in which the hoofs or toes are even in number, two or four on each foot. It includes the deer, cattle, sheep, goats, pigs, and related animals.

Key to the Families of ARTIODACTYLA

1a. Upper incisors present; canines tusk-like; muzzle pig-like. *TAYAS-SUIDAE,* page 198.

1b. Upper incisors absent; canines present or absent; muzzle not pig-
    like 2

2a. Antlers present, usually in males only (except *Rangifer*); antlers
    composed of solid bone, branched, shed yearly. *CERVIDAE*, page
    192.

2b. Horns present (not antlers) in both sexes; horns with a core of
    bone, covered with a horny sheath; horns not shed, or else not all
    of horn shed yearly...3

3a. Horns branched; all but bony core of horn is shed yearly. ANTI-
    LOCAPRIDAE, page 195.

3b. Horns not branched; no part of horn shed. BOVIDAE, page 196.

Family CERVIDAE
Deer and Relatives

1a. Size large, standing more than 4 feet high at shoulders.........4

1b. Size smaller, standing less than 4 feet high at shoulders. *Odocoi-
    leus* 2

2a. Tail brown above and white below, very conspicuous; tail longer,
    and used in a flag-like fashion for what appears to be signals;

metatarsal gland one inch long; antlers with one main branch and un-
branched tines; no white patch on
buttocks. **Fig. 377.** *Odocoileus vir-
ginianus.* **White-tailed Deer.**
Total length about 1800 mm.; tail
about 280 mm.; hind foot about 500
mm.; ear about 150 mm.; found in
the eastern and Rocky Mountain
states, and in a few localities in the
Pacific coast states.

White-tailed deer live mainly in
deciduous trees along streams and
in damp marshy regions, not inhabiting the coniferous forests to any extent. They eat mainly grass, all kinds
of leaves, and aquatic plants.

Figure 377

2b. Tail all black above, or black with a white tip or white with a
    black tip; tail shorter and not used as a flag; metatarsal gland
    2 to 4 inches long; antlers branch into 2 equal branches; white
    patch on buttocks as wide as tail or even wider................3

3a. Size about 1500 mm.; ear about 175 mm.; found in the humid coastal area of the west, but not in the dry interior mountains. Fig. 378. *Odocoileus hemionus columbianus*. Black-tailed Deer.

Figure 378

Tail about 175 mm.; hind foot about 450 mm. This deer is similar to the Mule Deer, but is smaller, has shorter ears, is darker in color, and has the metatarsal gland lower down in the hind legs than does the Mule Deer, and higher than does the White-tailed Deer. The tail is somewhat bushy, shorter than in the White-tail, and much wider than in the Mule Deer. This is the common deer west of the crest of the Cascade Range in Washington and Oregon, and occurs in the coastal areas of California as well.

3b. Size about 1750 mm.; ear about 250 mm.; found in the drier interior mountains of the west, but not in the humid coastal areas. Fig. 379. *Odocoileus hemionus hemionus*. Mule Deer.

Figure 379

Tail about 150 mm.; hind foot about 500 mm.; ears slender and long with a bushy black tip. The metatarsal gland is above the center of the metatarsal bone, is 4 or 5 inches in length, and is marked by long coarse hair. The Mule Deer is paler in color than is the Black-tailed Deer. Mule Deer are abundant on the ranges of the eastern parts of the Pacific coast states, in the Great Basin region, and in the Rocky Mountain states. They may be found in every forest where they are allowed to live, occurring in such large numbers in some regions that hunting must be opened for does as well as bucks.

4b. Antlers cylindrical.   Fig. 380.   *Cervus canadensis*.   Wapiti or American Elk.

Figure 380

Size about 2000 mm.; tail about 160 mm.; hind foot about 675 mm.; ear over 250 mm.; upper parts tawny with a conspicuous yellowish rump patch surrounding the tail; head, neck, breast, and legs dark brown; underparts reddish brown; neck with long shaggy hair in the males.

The Wapiti, or elk, the largest (except for the moose) of the North American deer, occur in the mountainous areas of the Rocky Mountain region. One subspecies occurs natively along the coast from the Olympic Mountains, Washington, south to northern California. They have been introduced in several other places on the western coast. The original range included a much greater section of North America than is now covered. During winter the elk leave their high feeding grounds in the mountain meadows and come down to less snowbound levels.

5a. Only males have antlers; size very large; no antler prong extending downward over the face. Fig. 381. *Alces alces*. Moose.

Figure 381

Total length about 2500 mm.; tail 65 mm.; weight 1000 pounds, or even more; color brownish black, but may be lighter.

The moose is not common in the United States, but does occur in the northern part of the Rocky Mountains, then northward into Canada. It can be seen in Yellowstone and Grand Tetons National Parks, and sparingly in Glacier Park. A bull moose is a massive, noble looking animal—at his best in a northern lake feeding on the aquatic plant life which is the main part of his food. Twigs and bark are also eaten.

5b. Both sexes with antlers; size smaller; antler with prong extending downward over face. Fig. 382. *Rangifer tarandus*. Caribou.

Size about 1800 mm.; tail about 100 mm.; weight about 250 to 500 pounds. Color blackish brown; underparts grayish white.

The caribou (often called reindeer) extends only a very short distance into the northern parts of Washington, Idaho, and Montana, but is common over most of Canada, adapting itself to barren northlands as no other of the deer family could.

Figure 382

Family ANTILOCAPRIDAE
Antelopes

The prong-horned antelope is not the true antelope of Africa. It is found only on the plains of the semiarid and arid western states. At one time it was almost depleted in numbers, but after careful protection for many years it has now come back until it can be considered again as one of the most interesting big game animals.

Fig. 383. *Antilocapra americana*. Antelope.

Total length about 1350 mm. to 1500 mm.; tail about 125 mm.; hind foot about 425 mm.; ear about 165 mm.; color tan with a conspicuous white rump, two white bands across the throat, a white area on the chest, and a white area along the side.

The antelope is unique in that it sheds the outer covering of its horns annually. Both sexes have horns. Another interesting characteristic is its ability, when excited, to erect the long white hairs on the rump to form a dazzling white circle that flashes for miles as the herd runs across the plains.

Figure 383

Family BOVIDAE
Bison, Sheep and Goats

Because everyone is familiar with the domestic cattle, sheep and goats, it is hardly necessary to describe this family. Members of the group occurred formerly in large numbers, but now they are nearly extinct in the United States.

Key to the BOVIDAE

1a. Size very large, cow-like; horns smooth and rounded, not greatly curved; size over 3000 mm.; tail 500 mm.; weight about 1800 pounds; color dark brown with a heavy mane in the males. Fig. 384. *Bison bison.* Bison or American Buffalo.

Figure 384

The Bison were formerly abundant over the prairie states, occurring in herds of hundreds of thousands. It is not known exactly how far west they ranged into the western states, but there is evidence that they extended well into eastern Oregon, eastern Washington, Idaho, northern California, northern Nevada, and northern Utah, as well as all the Rocky Mountain states. Now they may be found only in parks and zoos, although attempts to reestablish them in large areas like Yellowstone National Park have been rather successful.

1b. Size smaller, not cow-like; horns varying......................2

2a. Horns curved in a backward spiral; color grayish brown.   Fig. 385. *Ovis canadensis*.   Mountain Sheep or Big-horn Sheep.

Figure 385

Total length about 1500 mm.; tail about 50 to 100 mm.; hind foot about 400 mm.; color grayish brown with a large whitish rump patch; underparts somewhat lighter.

The mountain sheep or bighorn sheep occurs now in a number of localities in the Rocky Mountains from the Canadian Rockies (where it is most common) south through the Rockies of the United States, and into Mexico. It is also found in the desert mountain ranges of eastern Oregon, Nevada, Arizona, southern California and Baja California.

2b. Horns almost straight, black in color; color white.   Fig. 386. *Oreamnos americanus*.   Mountain Goat.

Figure 386

Total length about 1650 mm.; tail about 165 mm.; length of horns about 250 mm.; color white to yellowish white; horns and hoofs black; prominent chin whiskers in males.

These interesting antelopes (for that is what they really are) live only in the high rocky areas of the northern Rockies, the northern Cascades, the Olympic Mountains, then northward into Canada. They must have lived throughout the high mountains of the west at one time, and recent attempts have been made to reestablish them in some other areas of the west. Their sure footedness and agile grace in their precipitous, inaccessible mountain cliffs is a thing never to be forgotten by an observer.

Family TAYASSUIDAE
Peccaries

The peccary is the only wild pig in the United States. It is a small pig with three toes on the hind feet and four on the front feet, with a musk gland on top of the rump. The color is a mixture of black and white above, with a while band over the shoulders; underparts black.

Fig. 387. *Tayassu tajacu.* **Collared Peccary.**

Figure 387

Total length about 1000 mm.; tail about 15 mm.; hind foot about 200 mm.; weight about 45 pounds. Found in the arid regions of the Southwest, where it frequents dense thickets of mesquite and cactus. The animal is omnivorous, eating whatever it can find.

INDEX AND PICTURED GLOSSARY

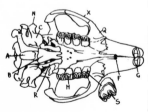
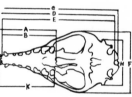

F

FAMILY: the systematic group below *order* but above *genus*; example, the cat family, Felidae, the squirrel family, Sciuridae.
FEMUR: the long hind leg bone nearest the pelvic bone.
FLANKS: the region of the hips, midway between dorsal and ventral sides.
FLUKES: the posterior appendage of a whale resembling tail fins of a fish.
FORAMEN MAGNUM: the large opening at back of skull through which the spinal cord passes. Fig. 391D

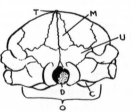

Figure 391

C — auditory bulla; D — foramen magnum; M—interparietal; O — occiput; T—sagittal crest; U—supraoccipital.

FOREARM (of bat: the longest bone of the arm (wing) to which the fingers are attached. Fig. 392

Figure 392

FRONTAL BONE: the bone just in front of the parietal and just behind the nasal.

G

GENERA: plural of *genus*.
GENUS: the systematic group just below *family* and above *species*; the genus is the first word of the scientific name of an animal or plant; example, Microtus (the meadow mice), Sciurus (tree squirrels), Felis (house cat).

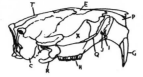

Figure 394

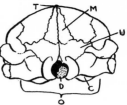

Figure 396

H

HERBIVOROUS: feeding upon plants.
Heteromyidae 114
HIBERNATION: the dormant state in which a mammal spends a period of time, such as in ground squirrels, and jumping mice, and during which time the body processes are greatly slowed down.

I

INCISIVE FORAMINA: the anterior palatine foramina (plural for foramen) of which there are two in the bony roof of the mouth in the anterior part at the junction of the premaxilla and maxillary bones; these foramina transmit the palatine arteries. Fig. 393F

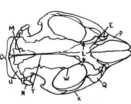

Figure 393

INCISORS: the very front teeth of a mammal. Fig. 394G
C — auditory bulla; E — frontal bone; G—incisors; H — molars; I — infraorbital canal; J—infraorbital foramen; P — premaxilla; Q — maxillary bone; R — pterygoid; T — sagittal crest; X — zygomatic arch.

INFRAORBITAL CANAL: a canal through the maxillary bone from the orbit of the eye to the face. Fig. 394I
INFRAORBITAL FORAMEN: the opening, in the infraorbital canal. Fig. 394J
INTERFEMORAL MEMBRANE: the fold of skin in a bat from the hind leg to the tail; the uropatagium.
INTERORBITAL BREADTH: the least distance across the skull between the orbital canal. Fig. 394J Fig. 395L

E — frontal bone; L — interorbital breadth; M — interparietal; N — mastoid; O—occiput; P — premaxilla; Q — maxillary bone; T — sagittal crest; U — supraoccipital; W — supraorbital process; X —zygomatic arch.
Interorbital constriction, see interorbital breadth.
INTERPARIETAL: the bone just in front of the supraoccipital, and between the two parietals or temporal bones. Fig. 396M

J

K

L

LACTATING: giving milk

M

MAMMAE: the mammary or milk glands.

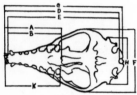

Figure 397

MAXILLARY ARCH: the an-
   terior portion of the
   zygomatic arch, really an
   extension of the maxillary bone.
MAXILLARY BREADTH:
   width of the skull across
   the maxillary bones. Fig.
   398G

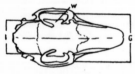

Figure 398

MAXILLARY MOLARS: the
   molars set into the maxillary bones.
MAXILLARY TOOTH ROW:
   the row of molars in the
   maxillary bone.

METACARPALS: the bones
   between the wrist and
   the fingers.
METATARSAL GLAND: the
   long gland on the hind
   foot of a deer.
METATARSALS: the bones
   between the ankle and
   the toes.
MOLARS: the posterior large
   teeth of a mammal.

MUZZLE: the long snout, including nose and mouth,
   of a mammal like a
   coyote or fox.

N

NAPE: the dorsal part of
   the neck just back of the
   head.
NOMENCLATURE: the sci-
   ence of naming things.
NOTCH (in the ear): the
   small indented spot at the
   base of the ear, opposite
   the tip; the ear measurement is taken from this
   notch to the opposite tip.
   Fig. 399

Figure 399

O

OCCIPUT: the back part of the skull. Fig. 400O

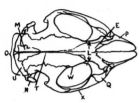

Figure 400

Ocelot 173
Ochotona princeps 70
Ochotonidae 69
Odobenus rosmarus 175
Odocoileus columbianus 193
    hemionus 193
    virginianus 192
OMNIVOROUS: feeding up-
    on almost anything, both
    plant and animal mater-
    ial.
Ondatra zibethicus 143
Onychomys leucogaster 136
    torridus 136
Opossum 43
ORDER: the systematic group
    below *class* and above
    family; example, the order
    Carnivora, or the order
    Rodentia.
Orders of mammals, key to
    39
Oreamnos americanus 197
Oryzomys palustris 140
Otter, America 163
    Sea 163
Ovis canadensis 197

P

PALATE: the bony roof of
    the mouth, composed of
    two palatine bones, two
    maxillary bones, and two
    premaxillary bones.
Paleontology 2
Panther 172
Parascalops breweri 48
PATAGIUM: the loose folds
    of skin along the belly
    of a flying squirrel by
    which the animal is able
    to glide or "fly."
Peccary 198
PECTORAL: the region of
    the chest or shoulders.
Perissodactyla 191
Perognathus alticola 117
    amplus 118
    apache 119
    baileyi 115
    californicus 116
    callistus 119
    fallax 116
    fasciatus 119
    flavescens 119

flavus 119
formosus 117
hispidus 115
inornatus 118
intermedius 116
longimembris 118
merriami 120
nelsoni 116
parvus 117
penicillatus 115
spinatus 115
xanthonotus 117
Peromyscus boylii 133
    californicus 132
    crinitus 134
    eremicus 133
    floridanus 131, 132
    gossypinus 135
    leucopus 135
    maniculatus 133
    nasutus 132
    nuttallii 135
    pectoralis 133
    polionotus 134
    taylori 134
    truei 132
PHALANGES: the bones of
    the fingers and toes.
Phenacomys albipes 146
    intermedius 145
    longicaudus 145
    silvicola 146
Phoca fasciata 177
    groenlandica 177
    hispida 178
    vitulina 179
Phocaena dalli 190
    phocaena 190
    vomerina 190
Phocaenoides dalli 190
Photography 33
Phyllostomidae 58
PHYLUM: the largest sys-
    tematic group of animals;
    example, the phylum
    Chordata (to which mam-
    mals belong), or the phy-
    lum Arthropoda (to which
    insects belong).
Physeter catodon 181
Pig, wild 198
Pigs, fossil 2
Pika 69, 70
Pine mouse 150
Pinnepedia 174
Pipistrellus hesperus 66
    subflavus 66
PLANTAR TUBERCLES: pro-
    tuberances on the sole of
    the foot.
PLANTIGRADE: walking flat
    on the sole of the foot.
Plecotus townsendii 65
    rafinesquii 65
Pocket gophers 108-114
Pocket mice
    Apache 119
    Arizona 118
    Bailey's 115
    Beautiful 119
    California 116
    Desert 115
    Great Basin 117
    Hispid 115
    Little 118
    Long-tailed 117
    Merriam's 120

Nelson's 116
Olive-backed 119
Plain's 119
Rock 116
San Diego 116
San Joaquin 118
Silky 119
Spiny 115
White-eared 117
Yellow-eared 117
Populations 9
Porcupines 156
Porpoises 189, 190
POSTAURICULAR
    PATCHES: areas of dis-
    tinct color just behind the
    ears.
Posterior process of supraor-
    bitals, see Supraorbital
    process.
POSTORBITAL: behind the
    eye.
POSTORBITAL SPINES: pro-
    jections of frontal or ju-
    gal bones behind the eye.
    Fig. 401W

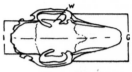

Figure 401

Prairie dogs 107, 108
PREHENSILE: a tail which
    curls around objects en-
    abling the animal to
    grasp with the tail.
PREMAXILLA: the bone
    bearing the incisor teeth,
    see figs. 400P, 404P.
PREMOLAR: smaller molar-
    like teeth just in front of
    molars.
Preserving in solutions 23
Primata 191
PROBOSCIS: a greatly elon-
    gated snout.
Procyon lotor 162
Procyonidae 162
Prodelphinus plagiodon 188
Pseudorca crassidens 186
PTERYGOIDS: the bones
    just behind the palatines.
    Fig. 402R
Puma 170

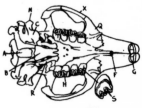

Figure 402

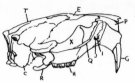

Figure 404

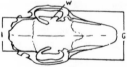

Figure 405

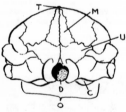

Figure 403

SAGITTAL CREST: the raised
  ridge of bone at the junction of the two parietal
  bones. Figs. 403T, 404T

T

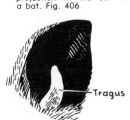

U

Figure 407

V

W

Z

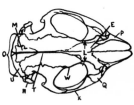

Figure 408